THE GENESIS OF FEYNMAN DIAGRAMS

Archimedes

NEW STUDIES IN THE HISTORY AND PHILOSOPHY OF SCIENCE AND TECHNOLOGY

VOLUME 26

Archimedes has three fundamental goals; to further the integration of the histories of science and technology with one another: to investigate the technical, social and practical histories of specific developments in science and technology; and finally, where possible and desirable, to bring the histories of science and technology into closer contact with the philosophy of science. To these ends, each volume will have its own theme and title and will be planned by one or more members of the Advisory Board in consultation with the editor. Although the volumes have specific themes, the series itself will not be limited to one or even to a few particular areas. Its subjects include any of the sciences, ranging from biology through physics, all aspects of technology, broadly construed, as well as historically-engaged philosophy of science or technology. Taken as a whole, *Archimedes* will be of interest to historians, philosophers, and scientists, as well as to those in business and industry who seek to understand how science and industry have come to be so strongly linked

For further volumes:
http://www.springer.com/series/5644

Adrian Wüthrich

The Genesis of Feynman Diagrams

Springer

Adrian Wüthrich
University of Bern
History and Philosophy of Science
Exact Sciences
Sidlerstrasse 5
3012 Bern
Switzerland
awuethr@philo.unibe.ch

ISSN 1385-0180
ISBN 978-90-481-9227-4 e-ISBN 978-90-481-9228-1
DOI 10.1007/978-90-481-9228-1
Springer Dordrecht Heidelberg London New York

Library of Congress Control Number: 2010934119

Printed on acid-free paper

Springer is part of Springer Science+Business Media (www.springer.com)

Acknowledgements

I wish to thank: *Gerd Graßhoff* for helping me turn my interest in Feynman diagrams into a research project in the University of Bern's Department of the History and Philosophy of Science, for supervising the project and for coaching me through the writing process; *Tilman Sauer* for co-refereeing this dissertation, for his many valuable comments and for his help during my period of research at the Caltech Archives, California Institute of Technology, Pasadena, California; the *Swiss National Science Foundation* for funding my research (research grant no. 100011–113589); The staff of the *Caltech Archives*, in particular *Judith Goodstein*, *Charlotte (Shelley) Erwin* and *Bonnie Ludt*, for granting me access to Feynman's notes, for helping me search and collate the material and for allowing such generous copying limits; *Diana Buchwald* and the *Einstein Papers Project* for their generous support during my period of research at the Caltech Archives; *Christian Wüthrich*, *Craig Callender* and their Reading Group at the University of California, San Diego for reading and discussing an early draft of some of my thoughts on Feynman diagrams; the *Quantum History Project*'s Reading Group, Department I at the Max Planck Institute for the History of Science, Berlin, for listening to and discussing a presentation of some of my early thoughts on Feynman diagrams, as well as *Jürgen Renn* for inviting me to be a guest in their department from April to August 2007; *Margareta Simons* for her careful and substantial copy-editing; *Christoph Greub* for helping me see how Theodore Welton's integral and differential form of equations relate to each other in detail; my parents, *Beatrice and Fritz Wüthrich*, as well as *Elsbeth Graf*, for their encouragement and support throughout my studies; my good friends for putting up with me being, far too often, "selectively antisocial"; *Magdalena González* for drawing the diagrams in section 4.6, as well as for giving me lots of moral support—badly needed at times.

Contents

List of Figures

Chapter 1
Introduction: Origin, Use and Interpretation of Feynman Diagrams

1.1 A Tool for Organizing Calculations

"Like the silicon chip of more recent years, the Feynman diagram was bringing computation to the masses." Thus Julian Schwinger,[1] displaying a hint of both disdain and admiration, appraises the enormous impact of Feynman diagrams as a mathematical tool on the daily work of theoretical physicists. And indeed, since Richard P. Feynman (1918–1988) invented them, around the year 1948, and Freeman J. Dyson subsequently systematized them, these diagrams have undeniably become an indispensable tool for performing calculations in modern quantum field theory. They are omnipresent in the theoretical treatments of an important class of elementary particle phenomena, in particular quantum electrodynamics (QED). In modern textbooks on quantum field theory—I will quote from one of them below—they take centre stage, while the teaching of their use is one of the essential components in courses on the subject.

Feynman diagrams enable one to find the relevant expressions in an approximate solution (perturbative expansion) to the equations that describe the dynamics of a quantum electrodynamic system. From the conception of the process as a sequence of particle creations and annihilations, one can produce a schematic drawing that relates these creation and annihilation events. The diagram thus obtained can then be translated, element by element, into a complex mathematical expression, the evaluation of which yields an observable quantity that characterizes the physical system (amplitudes of transition probabilities).

At the beginning of their widely used textbook on quantum field theory, Michael Peskin and Daniel Schroeder introduce Feynman diagrams in the following way:

> The main purpose of Part I of this book is to develop the basic calculational method of quantum field theory, the formalism of Feynman diagrams. We will then apply this formalism to computations in Quantum Electrodynamics, the quantum theory of electrons and photons. [...]
>
> Feynman diagrams provide for this elegant theory an equally elegant procedure for calculation: Imagine a process that can be carried out by electrons and photons, draw

[1] Schwinger (1983, p. 343).

A. Wüthrich, *The Genesis of Feynman Diagrams*, Archimedes 26,
DOI 10.1007/978-90-481-9228-1_1, © Springer Science+Business Media B.V. 2010

> a diagram, and then use the diagram to write the mathematical form of the quantum-mechanical amplitude for that process to occur. (Peskin and Schroeder 1995, p. 3)

We know of no methods that can exactly solve the equations that supposedly describe a quantum electrodynamic system, and no candidate for such an exact method even exists. Therefore, one has to turn to approximate solutions, such as the so-called perturbative expansion. For this method of solution Feynman provided an appropriate tool:

> The bad news is that even for this simplest of QED processes [$e^+e^- \rightarrow \mu^+\mu^-$], the exact expression for $\mathcal{M}$ [the quantum-mechanical amplitude for the process to occur] is not known. [...] The best we can do is obtain a formal expression for $\mathcal{M}$ as a perturbation series in the strength of the electromagnetic interaction, and evaluate the first few terms in this series.
>
> The good news is that Feynman has invented a beautiful way to organize and visualize the perturbation series: the method of *Feynman diagrams*. Roughly speaking, the diagrams display the flow of electrons and photons during the scattering process. (Peskin and Schroeder 1995, p. 5)

Besides organizing the calculations, Feynman diagrams also serve as a form of shorthand for the terms that occur in the calculations. Each element of a Feynman diagram corresponds to an elementary mathematical expression, while the Feynman diagram as a whole corresponds to the entire mathematical expression in which one is interested:

> According to the *Feynman rules*, each diagram can be translated directly into a contribution to $\mathcal{M}$. The rules assign a short algebraic factor to each element of a diagram, and the product of these factors gives the value of the corresponding term in the perturbation series. Getting the resulting expression for $\mathcal{M}$ into a form that is usable, however, can still be nontrivial. (Peskin and Schroeder 1995, p. 5)

As Peskin and Schroeder mention, the mathematical expression obtained by using the diagrams is still hard to evaluate, and the evaluation of such integrals can easily constitute a Masters or PhD thesis in theoretical physics.[2] Schwinger's analogy between the impact of the diagrams and that of the silicon chip should, therefore, be treated with caution. The power of Feynman diagrams to perform calculations is of a particular type: the diagrams do not aid us to evaluate integrals; rather, they help us to obtain the relevant terms from the unstructured and complex form of the approximate solution to the basic equations.

1.2 Tempting but Incorrect Interpretations

While Peskin and Schroeder (1995) aptly characterize Feynman diagrams as a tool for organizing perturbative calculations, Rom Harré (1988, pp. 62, 64) writes of a way of "parsing the amplitudes". To perform the "parsing" process, one has first to "imagine a process that can be carried out by electrons and photons", as Peskin

[2] See, for instance, Tödtli (2004).

and Schroeder state, and then draw diagrams that "display the flow of electrons and photons during the scattering process".

However, some textbooks explicitly warn the reader not to take too literally the interpretation that these diagrams represent scattering processes. The diagrams represent the scattering process only in an extremely abstract way. For most authors the safest way of using Feynman diagrams is, therefore, to consider them as nothing more than an abbreviation of the mathematical expressions that one has to deal with in the calculations:

> Please understand: these Feynman diagrams are purely symbolic; they do *not* represent particle trajectories (as you might see them in, say, a bubble chamber photograph). The vertical dimension is *time*, and horizontal spacings do *not* correspond to physical separations. For instance, in Bhabha scattering [see Fig. 1.1] the electron and positron are *attracted*, not repelled as the diverging lines might seem to suggest). All the diagram says is: "Once there was an electron and a positron; they exchanged a photon; then there was an electron and a positron again." Each Feynman diagram actually stands for a particular *number*, which can be calculated using the so-called *Feynman rules*. (Griffiths 1987, p. 59, emph. in the original)

F. A. Kaempffer's (1965) assessment of Feynman diagrams matches Peskin and Schroeder's introduction of the diagrams as an appropriate tool for keeping track of all the relevant terms in a perturbative expansion. And like David Griffiths quoted above, Kaempffer also emphasizes that it is easy to arrive at incorrect interpretations using the diagrams. He compares them to Faraday's field lines, which, according to Kaempffer, incorrectly suggested the existence of an ether.[3]

> After some contemplation of these drawings one is inevitably tempted to adopt, with Feynman, a picturesque manner of speaking about them which conjures up for every internal line the image of a particle *propagation* in a state of virtual existence between the vertices of that line. [...]
>
> One has to go back in the history of physics to Faraday's concept of the field line if one wants to find a mnemonic device which matches Feynman's graph in propagandistic persuasiveness. This historical analogy may serve here as timely warning against all too literal acceptance of mental images based mainly on a fabric of *conventions*, however consistent that fabric may appear. Thus, Maxwell was led by all too literal acceptance of Faraday's field concept to an ether theory of vacuum which ultimately turned out to be abortive. Similar temptations are lurking behind Feynman's graphs [...]

There are several kinds of arguments against giving Feynman diagrams a physical interpretation. Letitia Meynell (2008) classifies the arguments into two types of

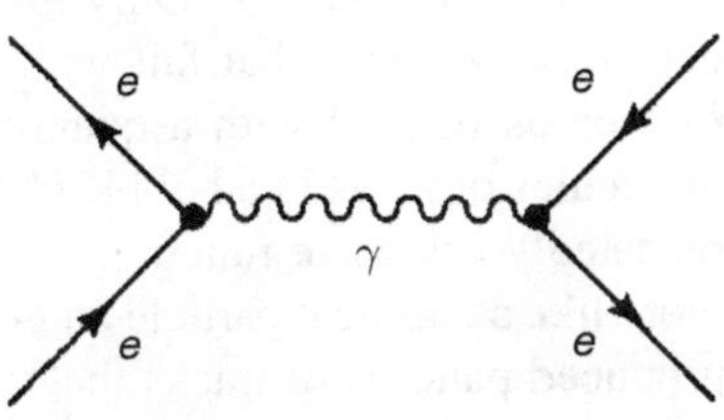

Fig. 1.1 One of the Feynman diagrams used in the discussion of Bhabha scattering (Griffiths 1987, p. 58)

[3] Kaempffer (1965, p. 209), emph. in the orginal; cf. Kaiser (2005, p. 369).

objections: the first objection is that Feynman diagrams represent something that, by definition, is unobservable, the so-called *virtual particles* referred to by Kaempffer in the above quotation; the second type of objection is that the concept of trajectories, which the lines in a Feynman diagram supposedly represent, is not, in fact, applicable in quantum mechanics.

The first type of objection is related to the question concerning the reality of the virtual particles. When a scattering process is described by the approximate solution (perturbative expansion), the terms in the expansion can be interpreted as the probability of transition from the initial to final states via various intermediate states. Most or all of these states, depending on the energy of the initial state, are unobservable because the energy of the initial state would not be enough to produce them. Nevertheless, the fact that, if sufficient energy were available, these states *could* be produced has to be taken into account when calculating the transition probabilities. However, one refers to intermediate states only because a particular method of solution has been chosen. Thus it is clear to Nick Huggett (2000, p. 629) that one should not attribute reality to these intermediate states: "[Virtual particles] are artifacts of an approximation scheme not particulate elements of reality."

Robert Weingard (1988) also states that virtual particles are not, as their name suggests, real after all. He points out that the intermediate states are *superpositions*, even of states with different numbers and types of particles. Therefore, the elements of the superposition do not represent real states of affairs as far as these characteristics are concerned. The situation is comparable to a double-slit experiment: a particle does not really pass through one or the other of the two narrow slits of a screen, neither does it pass through both of the slits, nor, indeed, through neither of them. Similarly, Weingard does not believe that virtual particles are really present in the intermediate states of a scattering process: "We do not have good grounds for thinking that Feynman diagrams picture real, albeit 'virtual', physical processes by means of which elementary particles interact" (p. 54).

In the Feynman diagram associated with a term in the perturbative expansion, the points of the creation and annihilation of a virtual particle are connected by an (internal) line, which might be taken to be the trajectory of that particle. The arguments to the effect that this is an incorrect interpretation belong to the second type of objections mentioned by Meynell (2008).

The concept of a trajectory is at variance with one of the fundamental principles of quantum mechanics. Only after measurements have been taken using particle detectors—devices that fall within the domain of classical physics—can a trajectory be associated with a quantum mechanical particle. The interaction with the molecules of a gas in a bubble chamber, for instance, causes the particle to localize repeatedly (its wave function collapses); the particle behaves, therefore, more and more like a classical particle and leaves an observable track. It is the experimentally produced patterns of tracks that may resemble Feynman diagrams. But one should refrain from interpreting the lines of the diagrams as the theoretical trajectory of the particles:

> Feynman diagrams look something like cloud chamber pictures, and they are often called space-time diagrams. This leads to the confusion. In fact, the diagrams do not picture physical processes at all. Instead, they represent probabilities (actually, probability amplitudes). The argument for this is very simple. In quantum mechanics (as normally understood), the Heisenberg uncertainty relations imply that no particle could have a position and a momentum simultaneously, which means there are no such things as trajectories, paths, through space-time. So the lines in a Feynman diagram cannot be representations of particles and their actual paths through space-time. (Brown 1996, pp. 265–267)

Thus, on the strength of such arguments, not only physicists such as Peskin, Schroeder, Griffiths and Kaempffer, but also historians and philosophers of science, such as James R. Brown, warn us against misinterpreting Feynman diagrams. Brown does not even acknowledge that they have an abstract representational function. He does not consider that they directly refer to any physical process in any way whatsoever. Rather, according to Brown, the diagrams are simply the abbreviation of complicated mathematical expressions and only the latter tell us something about the physical process:

> So what, then, is being visualized? I think the answer is simply this: Feynman diagrams are geometric representations of probability functions. [...] They are not pictures of phenomena. [...] [The] Feynman diagram geometrically represents (often brilliantly) a mathematical function which is linked to a physical process. We see the lines in the diagram; we do not visualize the physical process itself, nor any sort of abstract version of it. (Brown 1996, p. 267)

1.3 The Origin of Feynman Diagrams

The view that Feynman diagrams are simply a tool for organizing calculations and the many warnings against making incorrect interpretations also have a bearing on accounts of their origin. We already saw that Peskin and Schroeder (1995) consider that "Feynman has *invented* a beautiful way to organize and visualize the perturbation series" (my emph.). In the publication just cited, Brown also suggests that the diagrams are the result of Feynman's attempts to simplify the task of finding and organizing terms in complicated perturbative calculations:

> When Richard Feynman was working on quantum electrodynamics in the late 1940s, he created a set of diagrams to keep track of the monster calculations that were required. (Brown 1996, p. 265)

Some authors, such as S. S. Schweber and H. C. Baeyer, are more explicit in their claims that Feynman's preoccupations with complex mathematical expressions was the driving force behind the development of the diagrams:

> [Feynman] diagrams evolved as a shorthand to help Feynman translate his integral-over-path perturbative expansions into the expressions for transition matrix elements being calculated. (Schweber 1994, p. 434)

Baeyer, though, seems to distinguish between the diagrams' role as a form of shorthand and their role as a tool for organizing the calculations; he associates only the former to the context of their discovery:

> To minimize errors and save time, Feynman invented an ingenious graphic shorthand for the calculations, made up of little dots, lines and wiggles that can be thought of as plots of all the ways two or more electrons can interact. The drawings, now known universally as Feynman diagrams, look simple, but they stand for something very complicated. Feynman actually introduced each line, dot and wiggle not to represent a particle or the path of a particle, but to stand in for a complicated mathematical expression. Hence Feynman diagrams were initially a kind of stenography, but they ultimately made computations much faster and more reliable. Doing QED before Feynman was a bit like doing arithmetic with Roman numerals.[4]

Even Feynman himself retrospectively maintains that he made diagrams to help him "analyze perturbation theory quicker".[5] To the best of my knowledge, even in the retrospective account that most closely follows the development of the diagrams, Feynman recalls that they served mainly as a form of shorthand for mathematical expressions. The interview with Charles Weiner, to which I now refer, contains the following dialogue[6]:

Feynman: I can't tell you when I first wrote them. [...] I probably made diagrams to help me think about [perturbation expressions]. [...] It was probably not any specific invention but just a sort of a shorthand with which I was helping myself think, which gradually developed into specific rules for some diagrams. [...]

Weiner: For helping you think physically? In other words, you were seeing in physical—

Feynman: No, mathematical expressions. Mathematical expressions. A diagram to help write down the mathematical expressions.

Arthur I. Miller (1984) is one of the few historians who does not believe that Feynman diagrams were developed in order to find abbreviations for complex mathematical expressions. For him the development of Feynman's theory is more closely linked to physics than their function as a form of shorthand might suggest. Miller proposes two routes that might have led to Feynman's diagrammatic method. One is that Feynman continued his quest for *Anschaulichkeit* (visualizability) as in the later work of Werner Heisenberg on exchange forces in β decay.[7] Accordingly, he sees that one of the origins of Feynman diagrams lay in transforming diagrammatic representations of the exchange of quanta in *nuclear* physics. Miller emphasizes that this was not about finding illustrations more pleasing to the eye but rather about establishing different concepts of the physical phenomena:

---

[4] Baeyer (1999, p. 14); cited in Meynell (2008)

[5] Feynman (1966, p. 706).

[6] Weiner (1966a, pp. 41–42), reprinted with permission. Copyright 1966, American Institute of Physics.

[7] Miller (1984, e. g., p. 159).

> [The] transition from Wentzel's schematics of 1943 [(Wentzel 1943)] to the Feynman diagrams of 1949 required further transformations of what constitutes physical reality and its accompanying mental imagery. (Miller 1984, p. 166)

The other possible route Miller proposes is Feynman's attempt to interpret the divergences in QED by *renormalizing* the theory's parameters. However, Miller pursues neither of these two routes; he decides not to try and retrace the steps that led to the new and innovative representation of elementary particle phenomena because of "the complexity of the route" (p. 257).

In the most recent and one of the most comprehensive studies on Feynman diagrams, David Kaiser (2005) omits any discussion of their origin altogether. "[Kaiser's] project concerns instead what happened to the diagrams once they made the leap *out* of Feynman's head" (p. 5, emph. in the original). Kaiser describes how physicists in different places adopted and transformed Feynman diagrams, adapting them to their own uses so that the diagrams would satisfy their individual needs.

In most treatises on the subject, discussion on the origin of Feynman's method is clearly either cursory or entirely overlooked. And even Feynman's retrospective accounts are of little help: we learn from the interview with Weiner above that Feynman himself could not remember when he first used the diagrams.

1.4 The Reasons for the Diagrams' Success

When it comes to assessing the quality of the diagrams and finding reasons for their success, most accounts presuppose them to be a calculational tool with no representational function. Accordingly, weaker notions than "representation" have often been put forward to explain why Feynman diagrams have proved so popular with physicists, who clearly prefer them to alternative diagrams, such as the so-called dual diagrams.[8] The most prominent notions are those of *intuition* or *visualizability* and *similarity*, which are understood to share some but not all the characteristics of what constitutes a "representation".

1.4.1 Intuition and Visualizability

According to Miller (1984) the main appeal of Feynman diagrams is that they are "intuitive" or "anschaulich". In his discussion of them Miller refers to "visualizability", a concept that, according to him, is closely related but not identical to "intuition". Miller translates *Anschauung* as "intuition", *Anschaulichkeit* as "visualizability" and *anschaulich* as "intuitive" (p. 129). We encountered the verb "visualize" earlier in this section in the citations from Peskin and Schroeder and Brown, while Schweber (1986b) described "the power of visualization" and "impressive physical

8 The case of dual diagrams is discussed in Kaiser (2005, p. 366ff).

intuition" of Feynman's "style" as being the major characteristics of Feynman's "genius" (p. 504).

In the article featuring the first published Feynman diagram,[9] Dyson derives the results that Feynman found in a less sophisticated manner from the principles of quantum field theory. He tries to characterize the difference between Feynman's "intuitive considerations" (p. 496) and his more rigorous derivations as follows:

> In Feynman's theory the graph corresponding to a particular matrix element is regarded, not merely as an aid to calculation, but as a picture of the physical process which gives rise to that matrix element. For example, an electron line joining x_1 to x_2 represents the possible creation of an electron at x_1 and its annihilation at x_2, together with the possible creation of a positron at x_2 and its annihilation at x_1. (RadTh, p. 496)

Feynman described the dynamics of electrons and positrons in the framework of single-particle quantum mechanics. In particular he used the Green's functions associated with the Schrödinger and Dirac equations to describe the propagation of electrons forwards and backwards in time (see Chapter 5). On the basis of this theory and the interpretation of the diagrams as the abstract representation of particle propagation, he was led to find a type of diagram that, when translated into mathematical expressions, would contribute to a transition amplitude up to a given order of approximation.

Dyson derived the same quantitative results from the principles of quantum field theory and thus placed Feynman's results on a more established theoretical footing (see Chapter 6). He provided syntactical prescriptions for constructing the same diagrams that Feynman had developed on the basis of the latter's obsolete theory. Dyson also established a one-to-one-correspondence between the field theoretical quantities and these diagrams. Thus, once one had the diagrams, one also had the correct field theoretical expressions. The method by which one constructs the diagrams therefore becomes irrelevant. Therefore, one can use either Dyson's graphical prescriptions or, proceeding with due caution and restricting oneself to unproblematic cases only, Feynman's obsolete theoretical considerations. In this sense one can then obtain the correct quantum field theoretical expressions using Feynman diagrams "intuitively".

The problem with Feynman's theoretical assumptions is that they are based on equations for the *wave function* of electrons. By the 1940s it was generally agreed that quantum systems, in particular relativistic ones, should be described by mathematical field *operators* and not by functions. In other words, the theory which describes the system had to be quantized a second time.

Feynman's method was disqualified as being "intuitive" because of his obsolete theoretical assumptions. It was not the visual reasoning present in Feynman's papers that caused disagreement but the assumptions from which he derived his results (either using formulae or diagrams). In a lecture Dyson gave at Cornell University (Ithaca, NY)[10] 2 years after the publication of the article featuring the first published

[9] RadTh.

[10] Dyson (2006).

Feynman diagram, he is more explicit about his reasons for preferring his derivation to Feynman's[11]:

> The disadvantage of the Feynman theory is that it is constructed as a particle theory. The fact that there are many particles, indistinguishable from each other and obeying quantum statistics, has to be put into the theory as a special assumption. And the equations of motion of the particles become quite complicated when interactions between several particles, not to mention vacuum-polarization effects, are included. Thus the logical basis of the Feynman theory is much less simple than that of the field theory, where everything follows from general principles once the form of the Lagrangian is chosen.
>
> In this course we follow the pedestrian route of logical development, starting from the general principles of quantizations [sic] applied to covariant field equations, and deriving from these principles first the existence of particles and later the results of the Feynman theory. Feynman by the use of imagination and intuition was able to build a correct theory and get the right answers to problems much quicker than we can. It is safer and better for us to use the Feynman space-time pictures not as the basis for our calculations but only as a help in visualizing the formulae which we derive rigorously from the field-theory. In this way we have the advantages of the Feynman theory, its concreteness and its simplification of calculations, without its logical disadvantages.

Thus, on closer inspection, "intuition" as a concept turns out to concern the theoretical assumptions and not the means of representation used in deriving results from the theoretical assumptions. As such it cannot, therefore, explain the success of the diagrams.

1.4.2 Similarity

Another explanation for the diagrams' success is the notion of "similarity". Feynman diagrams closely resemble the tracks that particles leave in experimentalists' detectors; to some extent, they have the same syntactic elements as the photographs produced by bubble chambers and nuclear emulsions, experimentalists' principal detection devices. Harré (1988), whom I briefly mentioned in Section 1.2, believes that the close similarity of Feynman diagrams to particle tracks explains why they are physicists' preferred choice for representing the scattering processes described by quantum field theory.

Kaiser also singles out the similarity of Feynman diagrams to observed particle tracks as their distinguishing feature and compares them with an alternative diagrammatic method, the so-called *dual diagrams* (Kaiser 2005, p. 366ff.). In contrast to dual diagrams, which do not show this similarity, Feynman diagrams can be more easily related to the reality that is quantitatively described by the results of the calculations for which the diagrams are being used:

> The association of "realism" with Feynman diagrams in the 1950s and 1960s, based on their similarity to "real" photographs of "real" particles, helped Feynman diagrams stand out for many physicists. Unlike dual diagrams, Feynman diagrams could be read as more immediately related to real particles and processes, and hence less bound up with any

[11] Dyson (2006, p. 100); see also Sakurai (1967, pp. 240–241).

> particular abstract formalism. No one had to proclaim that Feynman diagrams were "the same" as nuclear emulsions, bubble-chamber photographs, or their stylized reconstructions for visual affiliations to be made. (Kaiser 2005, pp. 372–373)

However, as Nelson Goodman (1968) points out, it is doubtful whether similarity can be considered the basis for the realism of a representation. Which objects resemble other objects depends on which aspects of the phenomena are judged to be relevant. Along this line of thought it would be interesting to see how much of the realism that Feynman diagrams may have acquired and their claimed similarity to particle tracks are due to the appropriateness of the representation for the task of renormalization rather than to any independently established resemblance to the observable effects of the processes under study. Realism, as well as the perceived similarity, would then be a *consequence* of the diagrams' success, not a reason for it.[12] Their success would, in turn, then need to be explained by the appropriateness of a representation to a particular task.[13]

For all that, Kaiser leaves open the question of why "visual affiliations" to real particles should be advantageous to a calculational tool. Given the fact that the similarity between Feynman diagrams and the pictures of particle tracks incorrectly suggests that the diagrams represent trajectories, the associations that Feynman diagrams evoke appear to be more of a disadvantage than an advantage. Remember, for instance, Kaempffer's caveats quoted above: rather than exploiting the similarity between Feynman diagrams and real particle tracks, working physicists and teachers were busy trying to prevent colleagues and students from making incorrect interpretations on the basis of that similarity. They would probably have appreciated having at hand a less suggestive but equally powerful tool. Thus Feynman diagrams must have been adopted for other reasons.

1.5 The Ingredients of Modern QED

The theory of QED, as it stood in the 1930s and early 1940s, predicted the outcomes of experiments pertaining to it extremely accurately in those cases when the theory was applied in a rough approximation (perturbative expansion to first order). However, when physicists attempted to obtain yet more accurate results by applying QED in a more rigorous approximation (perturbative expansion to higher orders), the predictions became not only less precise but completely unusable and uninterpretable: they turned out to be infinite.[14]

To remedy this situation, the infinite quantities were parsed into finite and infinite parts, where the finite part could be argued as the physically relevant quantity. This implied that the infinite part could be discarded. One of the difficulties encountered

[12] Wüthrich (2007), cf. Goodman (1968).

[13] Cf. Graβhoff et al. (2002), Nickelsen (2002).

[14] For a comprehensive account of the history of QED, see Schweber (1994).

was finding a way to parse the quantities that was consistent with the theory of special relativity; in other words, to parse the quantities in a *covariant* way.

Because the calculations involve notorious subtractions of infinite quantities from infinite quantities, even a covariant equation could have solutions that were not covariant. A method was, therefore, needed in which each intermediate result was covariant. Such a method was proposed by Schwinger, Sin-Itiro Tomonaga and Feynman at the end of the 1940s, although Ernst C. G. Stueckelberg (cf. Pais 1986, p. 457f.) had, at least to some extent, put forward a method as early as 1934. The procedure proposed by these men to eliminate the infinities was the *renormalization* of the mass and charge of the electron.

1.5.1 Calculation Methods

Although he is one of its essential makers, Dyson does not regard the achievements of modern QED as genuine progress in physics. He sees modern QED and Feynman diagrams as a theory with the same physical assumptions as old QED; according to him, only the calculation methods changed. He writes that:

> Tomonaga, Schwinger, and Feynman rescued the theory without making any radical innovations. Their victory was a victory of conservatism. They kept the physical basis of the theory precisely as it had been laid down by Dirac, and only changed the mathematical superstructure. By polishing and refining with great skill the mathematical formalism, they were able to show that the theory does in fact give meaningful predictions for all observable quantities. (Dyson 1965, p. 589)

Max Dresden (1993, p. 55), however, believes that the transition from old to modern QED constitutes "a major technical advance with ramifications in many parts of physics", while Weinberg (1995, p. 37) describes the invention of Feynman diagrams as being of "practical importance". Feynman himself does not regard the invention of his diagrams as a discovery. In published articles on the subject, he sees them as a "mere re-expression of conventional quantum electrodynamics".[15] Indeed, later Feynman even admits to regretting having only reformulated already discovered theories.[16] He claims that the so-called vector-axial vector law of weak interactions was his only genuine discovery:

> It was the first time that I discovered a new law, rather than a more efficient method of calculating from someone else's theory (as I had done with the path-integral method for Schrödinger's equation and the diagram technique in quantum electrodynamics) [...] (Mehra 1994, p. 453)

Tomonaga, who shared the 1965 Nobel Prize in Physics with Feynman and Schwinger, relates in his Nobel Lecture that Sidney Dancoff correctly formulated all the assumptions from which the so-called *Lamb Shift* could have been

---

[15] STQED, p. 776.

[16] Feynman (1966, p. 707).

calculated.[17] So, had Dancoff not made a mistake in his calculations, the correct results for the Lamb Shift would have followed and, according to Tomonaga, "the history of renormalization theory would have been completely different". In the light of such events it seems plausible to conclude that suitable calculation methods were the only thing that prevented modern QED from being created.

Abraham Pais (1986) agrees that the developments in modern QED concerned mainly practical aspects. He also points out that such an assessment invites the question as to why the breakthrough did not come earlier:

> In later years, veterans of the 1930s battle with the infinite would occasionally and wistfully remark that this change should already have occurred in prewar days. They would remind themselves that, after all, the theory so masterfully developed further in the late forties was based on the same dynamical equations, subject to the same rules of quantum theory and relativity, with which they had struggled before. (Pais 1986, p. 455)

Some, like Victor Weisskopf, mention the work of the Swiss physicist Stueckelberg,[18] who seems to have accomplished much of what would only later be accepted as the correct method for treating the infinities:

> [It] seemed that a systematic theory could be developed in which these infinities were circumvented. At that time, nobody attempted to formulate such a theory, although it would have been possible then to develop what is now known as the method of renormalization.
>
> There was one tragic exception, and that was Ernst C. G. Stueckelberg [Stueckelberg 1934, 1938]. He wrote several papers in which a manifestly invariant formulation of field theory was put forward. This could have been a perfect basis for developing the ideas of renormalization. (Weisskopf 1983, pp. 73–74)

As Pais points out, "the delay (if one even wishes to call it like that)" can be partially explained by physicists' absorption in war-related work and their interest in nuclear physics.[19] Pais is not explicit about *what* exactly physicists failed to do. I assume that he means that physicists working in the 1940s simply did not have the time to redo and cross-check the huge amount of calculations that was required, lacking as they did the modern calculation method of Feynman diagrams.

1.5.2 Correct and Justified Predictions

Weinberg (1995, p. 38) writes that "the missing element was confidence in renormalization as a means of dealing with infinities", and puts forward the absence of trustworthy experimental data as the principal reason for this lack of confidence. J. Robert Oppenheimer, in his contribution to the Solvay Congress in 1948, that is

---

[17] Tomonaga (1973, p. 411); Dancoff (1939); Schweber (1994, p. 90); Pais (1986, p. 455); Brown (1993, p. 12).

[18] Ct. Lacki, Ruegg and Telegdi (1999).

[19] Pais (1986, pp. 455–456).

when QED was in transition, was of the same opinion. In order to explain why the Lamb shift was not correctly predicted earlier, he remarks:

> In their application to level shifts, these developments, which could have been carried out at any time during the last fifteen years, required the impetus of experiment to stimulate and verify. (Oppenheimer 1950, p. 271, reprinted in Schwinger 1958, p. 146)

Weinberg (1995, pp. 37–38) points out that experimental evidence for the level shifts mentioned by Oppenheimer was already available at the end of the 1930s.[20] However, the results only became reliable through the experiments of Willis Lamb and his collaborators.[21] The measured level shifts deviated from the values predicted by the old theory of QED that was based on Dirac's equation. Another deviation from Dirac's theory was encountered in the magnetic moment of the electron.[22] Using the renormalization of mass and charge, which characterizes modern QED, these values could now be accounted for.[23]

However, the fit between theoretical predictions and reliable experimental data does not completely justify the theory. Wolfgang Pauli, for one, is more concerned with other types of justification for renormalization:

> Even if one claims relativistic invariance and finiteness of the results this alone is not sufficient to make the subtraction-rules unique. One has to search for additional physical point of views. [...] The problem seems to be more to make the applied rules plausible and unique. (Pauli to Rabi, 15 January 1948, letter 931 in Pauli, Hermann and Meyenn 1979)

Dancoff (1939, p. 963), who was, according to the accounts mentioned above, only a calculational error away from the modern findings, tells the reader of "the fortuitous nature" of his results!

1.6 Representing and Calculating

That Feynman diagrams were characterized as being the chief tool in quantum electrodynamic calculations and that physicists warned others of incorrectly interpreting them as representing the trajectories of particles and, even worse, of particles in intermediate states, have been thoroughly examined. However, the inference from these observations that organizing calculations is the *only* function of the diagrams and that they cannot be consistently interpreted as representing physical processes is not sound. Diagrams can function simultaneously as idealized representations of the phenomena under study *and* as a tool for deriving statements about these phenomena.

Meynell (2008) elucidates the problem of representation in the case of Feynman diagrams by discussing accounts of different forms of representation mainly from

---

[20] For instance Pasternack (1938).

[21] See Lamb and Retherford (1947).

[22] Kusch and Foley (1948).

[23] Schwinger (1948a).

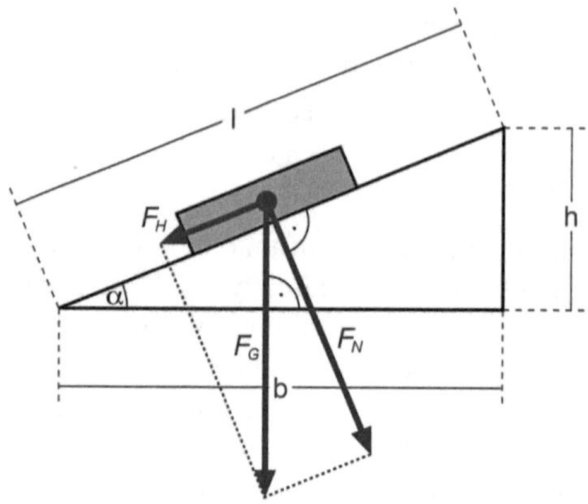

Fig. 1.2 Abstract drawing of a sliding body on an inclined plane. The diagram represents the relevant aspects of the physical situation and is used to determine quantitative relations between characteristic features of that situation. Source: `http://de.wikipedia.org/wiki/Kräfteparallelogramm`, last visited 22 Jan. 2009

the fields of aesthetics and the history of art, for example those of Goodman (1968) and Kendall Walton (1990). She also argues that the diagrams' function as a computational device does not preclude them from also being representations.

Familiar examples from the field of classical mechanics can act as cases in point, such as the graphical representation of a massive body on an inclined plane (see Fig. 1.2). Here, the abstract drawing serves to articulate the relevant aspects of the physical situation and, at the same time, to derive relationships between the vector of forces acting on the sliding object. As far as their principal functions are concerned, Feynman diagrams are no different.

Quantitative results are not always derived from the mathematical formulae obtained from the relations illustrated in the diagrams. At times the derivations are conducted directly by means of the diagrams. In her study on Venn diagrams, Sun-Joo Shin (1994) writes that diagrams can provide the means by which statements can be derived from other statements just as well as formulae. The representation of positrons in an early stage of the development of Feynman diagrams is an example of a proof conducted, to a large extent, directly by diagrams (see Section 4.6.2) in which Feynman performs what I like to call a *diagrammatic induction*. For another instance of diagrammatic induction, using a more familiar example, see Appendix A.

1.6.1 Developing Appropriate Means of Representation

If we acknowledge that diagrams can simultaneously fulfil the two principal functions of representing and facilitating derivations, then the different steps in the development and systematization process of Feynman diagrams can be better understood than if we presuppose that they serve only as a calculational tool.

Throughout the gradual process of their development, Feynman diagrams always fulfilled the dual function of representing the phenomena in an idealized manner and facilitating derivations of propositions concerning the idealization. In the early stages of their development (Section 4), the diagrams represented the motion of an electron in an abstract way. At the same time, Feynman used them to derive the quantitative characteristics of the electron's (and positron's) motion. The inconsistencies Feynman encountered and his desire to generalize the theory to interacting particles led him to modify his means of representation repeatedly. They grew more and more abstract, and representational elements of propagating electromagnetic potentials in classical electrodynamics were also incorporated (see Chapter 5). The interpretation of the graphical elements changed as the theoretical background evolved in parallel. But the diagrams always referred to relevant aspects of the phenomena *and* were used to derive quantitative results or other theoretical statements.

Appropriate means of representation are not restricted to diagrams, although the latter are the focus of my study. In Section 3 we will see that Feynman was already preoccupied with devising alternative descriptions to obtain particular advantages while working on his PhD thesis, in which the alternative representations concerned mainly the mathematical formulation of the theory of quantum systems. The main advantage of Feynman's alternative mathematical formulation of non-relativistic quantum mechanics is that it could be generalized to cases that had hitherto been indescribable.

In my account of the origin of Feynman diagrams, Feynman's preoccupation with complicated mathematical expressions thus recedes into the background, while his efforts to develop an appropriate way of representing quantum electrodynamic phenomena comes to the fore. From the time of his PhD thesis (defended in May 1942) to the publication of STQED (1949), Feynman was often absorbed in devising new means of representation with particular advantages.

Schwinger's work also largely consists of finding an appropriate means of representation. In contrast to Feynman, Schwinger did not (at least not in his published work) use diagrams but rather kept to mathematical formulations. But mathematical formulations, too, are a means of representing physical phenomena. Schwinger's main achievement was to find a mathematical formulation which showed, at every step of the calculation, the relativistic covariance of the observables.[24] To use such an explicitly relativistic formulation was not only important because one was dealing with infinite quantities; it was also necessary for identifying the terms that could be interpreted as being contributions to the mass of the electron. Only relativistic invariant terms allow for such an interpretation.[25]

---

[24] Schwinger (1948b, abstract)

[25] See, e. g., Feynman's letter to Mr and Mrs Corben, cited in Schweber (1994, p. 426): "Actually, the self-energy comes out finite and invariant and is therefore representable as a pure mass."

1.6.2 Features of Appropriateness

As previously mentioned, Weinberg describes the invention of Feynman diagrams as being of "great practical importance". But even though it may, therefore, appear that Feynman diagrams only simplify the application of the theory, Weinberg also points to the specificity of the simplification that Feynman diagrams provide: they organize the terms in the calculations not just in a way which assures that every term is taken into account but in more important theoretical respects[26]:

> One result of great practical importance that came out of Feynman's work was a set of graphical rules for calculating S-matrix elements to any desired order of perturbation theory. Unlike the old perturbation theory of the 1920s and 1930s, these Feynman rules automatically lumped together particle creation and antiparticles annihilation processes, and thereby gave results that were Lorentz-invariant at every stage. We have already seen in Weisskopf's early calculation of the electron self-energy [Weisskopf (1934)], that it is only in such calculations, including particles and antiparticles on the same footing, that the nature of the infinities becomes transparent.

Similarly, Dyson also acknowledges in SM that, although "using no new ideas or techniques" (p. 1754), he introduced

> into the theory what is really a new physical hypothesis, namely that the electron–positron field always acts as a unit and not as a combination of two separate fields. A similar hypothesis is made for the electromagnetic field, namely that this field also acts as a unit and not as a sum of one part representing photon emission and another part representing photon absorption. (SM, p. 1754)

Clearly, one cannot claim that Feynman diagrams alone provide a theoretical justification for the renormalization procedure. Nor could I sufficiently investigate Pauli's opinion of Feynman diagrams. But if the phenomena are analyzed using Feynman diagrams, the corresponding mathematical formulation gave Dyson the uniqueness of the "subtraction rules" involved in renormalization that Pauli insisted was necessary:

> Therefore it may be concluded that the rules of calculation of $U(\infty)$ are not only divergence-free but unambiguous. (SM, p. 1753).

Dyson (RadTh, SM) was able to isolate the problematic divergences using the diagrams. He identified basic graphical elements, the repeated insertion of which into a diagram that may already contain such elements describes the processes that are responsible for the infinite values in quantitative predictions. The way that quantum electrodynamic phenomena are represented in Feynman diagrams in turn suggests an appropriate mathematical formulation of the calculations, in which "the well-known divergences seem to have conspired to eliminate themselves".[27]

The specific appropriateness of this mathematical formulation explains, at least in part, both the "delay" in the creation of modern QED and the success of the

---

[26] Weinberg (1995, p. 37); cf. Sakurai (1967, p. 240).

[27] SM, p. 1754.

diagrams: modern QED could not be created as long as the formulation that came with the diagrams did not exist.

1.7 Inconspicuous Conceptual Innovations

As we saw in the quotation from Dyson above, the development of appropriate calculation methods can bring with it new physical hypotheses. Apart from the hypothesis mentioned by Dyson, a new concept of the charge and mass of an electron was created during the quest to eliminate the uninterpretable infinities from QED. Despite its historical origin, renormalization consists of more than just a method for removing uninterpretable infinite quantities. Even in a completely convergent formalism of QED (if such were to exist) a renormalization of masses and charges would be necessary.

Take an electron, for example. In its vicinity (virtual) electron–positron pairs are created which screen the original electron's charge (see Fig. 1.3). The "real", "bare" charge of the electron is thereby reduced to an "effective" charge, which has the observed value. In QED the bare charge is infinite and so is the strength of the screening. The two infinite values are such that their difference (or ratio, depending on the details of the renormalization method) yields the observed value of the charge. However, this renormalization of the charge would also be necessary if the bare charge and the screening were not infinite.

The essential lesson we need to learn from the renormalization method is that, for instance, an electron cannot be considered in complete isolation but is always accompanied by virtual electron–positron pairs and photons. Thus one of the implicit assumptions responsible for the uninterpretable infinities has to be revised. For example, in 1939 Weisskopf formulated the problem of the infinite self-energy of the electron with this flawed assumption in the background:

> The self-energy of the electron is its total energy in free space when isolated from other particles or light quanta. (Weisskopf 1939, p. 72)

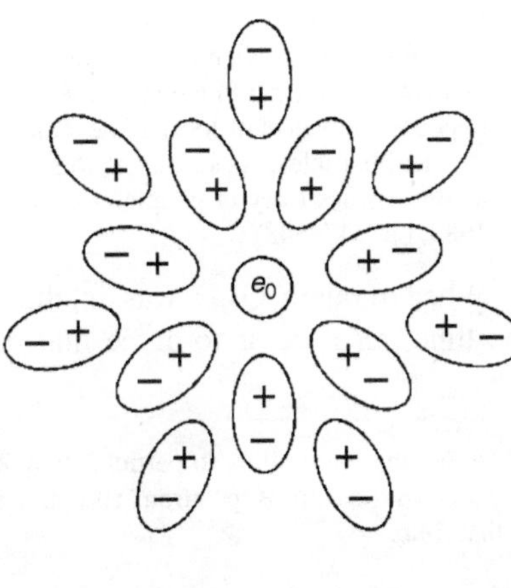

Fig. 1.3 Screening of the electron's charge by an assembly of virtual electron–positron pairs (Peskin and Schroeder 1995, p. 255)

The isolated electron is a meaningless concept since its very existence implies the creation of virtual photons, electrons and positrons. One has, therefore, to incorporate these effects into the definition of a "free" electron and distinguish between the charge parameter that appears in the theory's equations of motion (bare charge) of the hypothetical isolated electron and the observed value of the charge. The observed value is the difference between (or ratio of) the bare charge and the screening.

The same is true for *mass* renormalization. In renormalized QED it is understood that the mass of the electron is the joint effect of a bare mechanical mass and the coupling processes between the electron–positron field and the electromagnetic field.

Schwinger is one of the few actors who, in retrospect, would see significant conceptual advances in the development of renormalization. He emphasizes the aforementioned fact that renormalization is not, in effect, connected to the problematic infinities, and that the development of modern QED required a physical understanding of the origin of these infinities. However, Schwinger does not attribute the conceptual advances to Feynman diagrams; according to him, they concern only the superficial calculation methods:

> Topology (Feynman diagrams) is optional here; that is a matter of pedagogy, not physics. [...]
>
> What I have just described is all technique. Now, here is the music. It is probably a fairly wide-spread opinion that renormalized quantum electrodynamics is just the old quantized version of the combined Maxwell and Dirac equations, with some rules for hiding divergences. That is simply not true. A theory has two aspects. One is a set of equations relating various symbols. The other is, at some level, the physical interpretation to be associated with the symbols. In the course of the development here being described, the equations did not change, but the interpretation did. In the late 1930s, most people would not have challenged these statements: e and m, as they enter the Dirac and Maxwell equations, are the charge and the mass of the electron; an electromagnetic field operator creates or annihilates a photon; a Dirac field operator creates an electron or annihilates a positron, and its adjoint field does the inverse. And all this would be true if the two fields were uncoupled. But, in the real world, the localized excitation represented by an electromagnetic field, for example, does not just create a photon; it transfers energy, momentum and angular momentum, and then Nature goes to work. And so, it may create a photon, or an electron–positron pair, or anything else with the right quantum numbers.[28] The various Green's functions are the correlation functions among such localized excitations, and the study of their space-time behavior is the instrument for the identification of the physical particles and of their interactions. Renormalization, properly understood, is an aspect of the transfer of attention from the initial hypothetical world of localized excitations and interactions to the observable world of the physical particles. As such, it is logically independent of divergences. Could we construct a convergent theory of coupled fields, it would still need to be renormalized. (Schwinger 1983, pp. 347–348)

When modern QED was in the making, Schwinger, in published articles of the time, also seems to agree that he and other physicists did not come close to

[28] Here Schwinger refers to particles like mesons which were not yet incorporated into a theory together with electrons, positrons and photons at the end of the 1940s; or even not yet discovered at that time.

understanding the basic assumptions of the theory; this would have required "the introduction of fundamentally new concepts" (Schwinger 1948b, abstract). I have already quoted Dyson and Pais, who made similar comments (see Section 1.5.1).

In the same vein Peter Galison (1983) concludes that the attempted revolution against old QED "failed". He sees the main reason for its failure in the discovery of the muon, a particle similar to the electron but about 200 times heavier. Galison believes that the doubts shed on the validity of QED by some of the properties of cosmic radiation then began to dissolve. The need to revise the basic physical assumptions of the theory seemed to disappear after the discovery of the muon.

Weinberg (1977, p. 17) considers that the development of modern QED constitutes a strange type of progress:

> [There] seems to be general agreement that the essential element of scientific progress is a decision to break with the past.
>
> I would not quarrel with this view, as applied to many of the major advances in the history of science. It certainly seems to apply to the great revolutions in physics in this century: the development of special relativity and of quantum mechanics. However, the development of quantum field theory since 1930 provides a curious counterexample, in which the essential element of progress has been the realization, again and again, that a revolution is unnecessary. If quantum mechanics and relativity were revolutions in the sense of the French Revolution of 1789 or the Russian Revolution of 1917, then quantum field theory is more of the order of the Glorious Revolution of 1688: things changed only just enough so that they could stay the same.

1.8 A comparison with Traditional Means of Representations

If one then compares Feynman diagrams with the diagrams used by authors who did not yet know of them, one can see how much modern QED and the conceptualization and representation of quantum electrodynamic phenomena changed with the development and introduction of what appears to be a mere method of calculation. The two cases I will discuss (in Section 2) are the treatment of the scattering of light off light by Hans Euler (1936) and the renormalization methods proposed by the Japanese physicists working around Tomonaga.[29]

Euler did not know of Feynman diagrams simply because they had not yet been invented, whereas the Japanese physicists knew almost nothing about them because of their isolation from the community of physicists working in Europe and the United States in the aftermath of the Second World War.

The conspicuous feature of Euler's and the Japanese physicists' treatment of the problems of QED is that they relied on modified atomic level schemes. Their graphical representations display the energy levels of the electron and the positron, between which the particles made transitions by absorbing or emitting light quanta. These physicists used these diagrams, as well as long tables in Euler's case, to

[29] RadReacI; RadReacII; RadReacIIIa; RadReacIIIb.

organize the complicated calculations, much in the same way that one would later apply Feynman diagrams.

Feynman diagrams and the representations that Euler and the Japanese physicists used also have several features in common. In both cases the elements of the approximate solution to the basic equations (the perturbative expansion) are interpreted as being the most relevant physical aspects of the real process. And in both cases the real process was thus parsed into elementary processes consisting of the emission and the absorption of a photon by an electron or a positron.

However, Feynman diagrams analyze the phenomena in elementary processes in a way that shows the renormalization procedure to be the incorporation of the net effect of Schwinger's "music" and the mass and charge of the hypothetical, isolated electron. The unrenormalized description of the phenomena would correspond to a fractal-like Feynman diagram into which ever more elementary parts are inserted (cf. Fig. 6.3). These insertions represent the plethora of possible creation and annihilation processes. The mass and charge parameters of the theory are calibrated in the renormalization procedure so that these inevitable processes are included in the inertial motion and unit interaction with the observed mass and charge.

Although term schemes may well have been the most appropriate form of representation when it came to a traditional interpretation of the equations, Feynman diagrams were more appropriate to the modern conception. For example, they showed whether the same photon is involved in an emission or absorption process, or whether two different photons are emitted or absorbed (see Section 2.3). This is particularly relevant when it comes to finding a divergence-free, empirically adequate QED, but irrelevant in the original context of term schemes that represent the energy levels of atomic electrons in a surrounding electromagnetic field, conceived of as a reservoir of photons.

The conceptual framework that Euler and the Japanese physicists presupposed by using these graphical representations differs from that adopted by physicists using Feynman diagrams, and so they would probably not have understood Peskin and Schroeder's later instructions (see Section 1.1). Physicists working with (modified) term schemes considered "a process" to be a sequence of transitions between the energy levels of electrons and positrons caused by the emission and absorption of quanta of the surrounding electromagnetic field, which is reminiscent of atomic physics; physicists working with Feynman diagrams, however, see "a process" as the interruption, by the emission or absorption of a photon, of the electron's or positron's free propagation, which is described by quantum field theoretical vacuum expectation values. The statement "Imagine a process that can be carried out by electrons and photons" differs, therefore, in meaning, depending on whether one is working with Feynman diagrams or with the modified term schemes.

Feynman diagrams replaced the (modified) term schemes that were used in the early treatment of quantum electrodynamic phenomena both as a means of representation of the phenomena and as a calculational tool. With hindsight, the development of the former is easily overlooked. Schweber (1994, pp. 119, 122–123), for instance, represents the processes to which he refers in his account of the theoretical development of QED in the 1930s by Feynman diagrams, even though, at that time, they had not even been invented. Schweber does not and indeed, therefore, cannot identify the conceptual advances from the old form of QED to the new theory of QED, which the development of Feynman diagrams brought about. Rather, it becomes clear that physicists like him[30] had become so used to representing QED phenomena through Feynman diagrams that the conceptualization they adopted appeared trivial to them and thus unworthy of mention.

Whether the term "revolution", which Thomas Kuhn (1962) prominently introduced into the discipline of the history and philosophy of science, adequately characterizes the development of modern QED is not my concern here. Significant advancements did take place and "no revolution could have done more".[31] If the transition from the old theory to the modern theory of QED is to be labelled a revolution at all, it should not be described as a conservative, purely technical or even failed revolution, as most accounts suggest; rather, it was a *creeping* revolution: a series of inconspicuous but fundamental revisions to the foundations of QED that can be unravelled by following the development and systematization of Feynman diagrams and then comparing them with the traditional means of representing quantum electrodynamic phenomena.

[30] Although, since some time, mainly working in the field of the history of physics, Schweber was long enough a practising physicist and made notable contributions also to this discipline (Bethe and Schweber 1955, to mention but one of them).

[31] Dresden (1993, p. 55).

Chapter 2
Quantum Electrodynamics Without Feynman Diagrams

If one compares Feynman diagrams with other diagrammatic representations of the same phenomena, one can discern the change in the conception and representation of the phenomena that silently occurred while Feynman diagrams were being developed by Feynman and then systematized by Freeman J. Dyson. In some publications, diagrammatic representations occupy centre stage, which testifies to their importance in the context of quantum electrodynamics (QED). In this section we will see that some authors even explicitly mention that improving diagrammatic representations was one of their principal goals.

The two cases that I will discuss in this section are the diagrams used by Hans Euler (1936) and by Ziro Koba and Gyo Takeda (1948/49a, b).[1] In the first case Feynman diagrams had simply not been invented. In the second case, the Japanese authors were unaware of Feynman diagrams since they were isolated from the European and American scientific communities in the aftermath of the Second World War. What both cases have in common, however, is the fact that the diagrams are modifications of the term schemes used in atomic physics. The Japanese physicists adopted the diagrams from Euler and then modified them further.

Euler employed modified term schemes in order to see how the classically impossible scattering of light by light could come about, while Koba and Takeda, much like Dyson (see Chapter 6), employed their version of the term schemes to show that the various infinite terms occurring in the calculations cancel each other out.

2.1 Scattering: The Transition Between Energy Levels

From Halpern (1933) at the very latest, it was known that, in the absence of any material obstacles, a light quantum could be scattered by another light quantum, in other words that the "vacuum" has "scattering properties".[2] Maxwell's equations, which describe the propagation of light, are, therefore, not entirely accurate. According to these equations, two light waves superpose without

[1] RadReacI; RadReacII; RadReacIIIb; RadReacIIIa.

[2] Halpern (1933, p. 856.)

A. Wüthrich, *The Genesis of Feynman Diagrams*, Archimedes 26,
DOI 10.1007/978-90-481-9228-1_2, © Springer Science+Business Media B.V. 2010

interacting; as such the scattering of light by light would, therefore, correspond to the introduction of a non-linear term.

Compared with traditional electrodynamic and radiation problems examined at the time, light-by-light scattering was also untypical, as it involves the quanta of the radiation field and not material particles, such as electrons. Term schemes are particularly suited to the latter case, since they represent the energy levels of the material system as horizontal lines and the quanta, at best, as vertical lines or arrows indicating the absorption or emission of a photon. Photons were, above all, a means by which one could probe the structure of matter.

Despite the fact that light-by-light scattering is untypical of traditional radiation problems, Euler (1936) employs a traditional means of representation—term schemes—in his theoretical treatment of it (see Fig. 2.1), which indicates that, at that time, there was no obvious alternative.

Hans Euler and Otto Halpern explain light-by-light scattering in the framework of Dirac's hole theory. Each of the two incident light quanta can, for instance, lift an electron out of a negative energy state into a positive energy state, thereby leaving a hole in all the negative energy states previously occupied by the electrons. This process is observed as the formation of an electron–positron pair, the hole being interpreted as the positron. The electron–positron pair can then annihilate again to radiate a light quantum. Light-by-light scattering occurs if the two intermediate events of pair creation and annihilation do not occur independently but as the result

Lichtquanten \ Elektronen	Absorption von g^1	Absorption von g^2	Emission von $-g^3$	Emission von $-g^4$
$\mu = 1$	Paarerzeugung	Elektronensprung	Elektronensprung	Paarvernichtung
$\mu = 2$	Paarerzeugung	Elektronensprung	Positronensprung	Paarvernichtung
$\mu = 3$	Paarerzeugung	Positronensprung	Elektronensprung	Paarvernichtung
$\mu = 4$	Paarerzeugung	Positronensprung	Positronensprung	Paarvernichtung
$\mu = 5$	Paarerzeugung	Paarerzeugung	Zerstrahlung des Elektrons vom 1. mit dem Positron vom 2. Paar	Zerstrahlung des Positrons vom 1. mit dem Elektron vom 2. Paar
$\mu = 6$	Paarerzeugung	Paarerzeugung	Zerstrahlung des Positrons vom 1. mit dem Elektron vom 2. Paar	Zerstrahlung des Elektrons vom 1. mit dem Positron vom 2. Paar

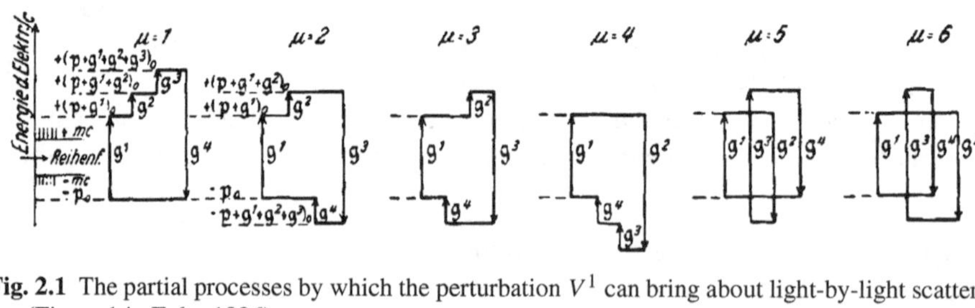

Fig. 2.1 The partial processes by which the perturbation V^1 can bring about light-by-light scattering (Figure 1 in Euler 1936)

of the electron produced by one photon annihilating the positron produced by the other photon.

Light-by-light scattering can even occur if the energy of the two incident photons does not equal the energy of the two electron–positron pairs that have to be produced for the scattering to take place. The hypothetical possibility of the creation of the pairs does affect the cross section of the scattering, even if the possibility does not actualize. This is comparable to the effect of the second slit in the double-slit experiment: whether the second slit is open or not, that is, whether the electron can pass through the second slit or not, affects the probability of the electron being detected after having passed through the first slit.

The case where the energy suffices to produce "real" pairs can be treated by the formulae provided by Gregory Breit and John A. Wheeler (1934). Euler provides an adequate description and quantitative evaluation for the case in which the pairs are only "virtual"[3]:

> *Either:* the energies cg^1 and cg^2 of the two light quanta and the angle enclosed by their momenta $\mathfrak{g}^1$, $\mathfrak{g}^2$ are so large that the energy and the principle of momentum conservation allow the creation of a *real* pair $(g^1g^2 - (\mathfrak{g}^1\mathfrak{g}^2) > 2(mc)^2)$. Then, one obtains the scattering probability of the light quanta off each other by multiplying the probabilities of pair creation and annihilation and summing over all the possibilities. This has been done by Breit and Wheeler.
>
> *Or:* the energy and momentum of the two light quanta do not suffice to create a real pair $(g^1g^2 - (\mathfrak{g}^1\mathfrak{g}^2) < 2(mc)^2$, i. e. in an appropriate reference system: $g^1 < mc$, $g^2 < mc)$. Then, the light quanta g^1, g^2 may, because of the *virtual* possibility of pair creation, nevertheless evolve into two light quanta, and a scattering of light by light must also occur in this case (for instance, visible light). (Euler 1936, p. 399, emphasis in the original)

The interaction between the light quanta and the electrons is described in the first-order perturbation theory by the term

$$V^1 = e \int \psi^*(\alpha\mathfrak{A})\psi \, dV, \tag{2.1}$$

where e is the charge of an electron, ψ and ψ^* the electron's (spinor) wave function and its complex conjugate, α the vector of Dirac matrices, $\mathfrak{A}$ the electromagnetic vector potential and dV the volume element for integration.[4]

---

[3] My translation of the original German text:
"*Entweder* die Energien cg^1 und cg^2 der beiden Lichtquanten und der Winkel zwischen ihren Impulsen $\mathfrak{g}^1$, $\mathfrak{g}^2$ sind so groß, daß Energie und Impulssatz die Erzeugung eines *wirklichen* Paars erlauben $(g^1g^2 - (\mathfrak{g}^1\mathfrak{g}^2) > 2(mc)^2)$. Dann erhält man die Wahrscheinlichkeit der Streuung der Lichtquanten aneinander, indem man die Wahrscheinlichkeiten der Paarerzeugung und der Wiederzerstrahlung multipliziert und über alle Möglichkeiten summiert. Dies ist von Breit und Wheeler durchgeführt worden.

Oder aber Energie und Impuls zweier Lichtquanten reichen nicht zur Erzeugung eines wirklichen Paares aus $(g^1g^2 - (\mathfrak{g}^1\mathfrak{g}^2) < 2(mc)^2$, d. h. in geeignetem Bezugssystem: $g^1 < mc$, $g^2 < mc)$. Dann können die Lichtquanten g^1, g^2 doch durch die *virtuelle* Möglichkeit der Paarerzeugung in zwei Lichtquanten übergehen und auch in diesem Fall (etwa des sichtbaren Lichts) muss es eine Streuung von Licht an Licht geben."

[4] Euler (1936, pp. 416/417).

Such a term describes the annihilation of an electron–positron pair into a photon or the creation of an electron–positron pair by the absorption of a photon. It does not directly describe the transition of a state with two light quanta into a state with, again, two light quanta, for which the energy is insufficient. However, since in the transitions to intermediate states energy does not have to be conserved, the interaction term above can, nevertheless, describe the scattering of light by light, but only in the *fourth* order. A term, V_{in}^4, introduced by Werner Heisenberg[5] to circumvent divergence difficulties does describe the transition from two light quanta (state i) to two light quanta (state n) directly (even at low energy) and also needs to be taken into account. The matrix element, H_{in}^4, describing low energy light-by-light scattering is, therefore,[6]

$$H_{in}^4 = \sum_{kem} \frac{V_{ik}^1 V_{ke}^1 V_{em}^1 V_{mn}^1}{(E_i - E_k)(E_i - E_e)(E_i - E_m)} + V_{in}^4, \tag{2.2}$$

where the Es denote the energies of the intermediate states k, e and m.

Euler systematically lists the various possible sequences of "partial processes" by which the perturbation V^1 can bring about light-by-light scattering in a table complemented by a series of diagrams (see Fig. 2.1).

There are six different possibilities that can explain the behaviour of the intermediate electrons and positrons. In every case an electron–positron pair has to be created first and the last partial process has to be the annihilation of an electron and a positron. Therefore, the six possibilities concern the behaviour of the electrons and positrons between these two processes. The first possibility considered ($\mu = 1$) is: initial creation (*Paarerzeugung*), then two changes of energy of the created electron (*Elektronensprung*) through the emission or the absorption of a photon, and finally the annihilation of the pair (*Paarvernichtung*). The second possibility ($\mu = 2$) is that the penultimate partial process is not the change of energy of the intermediate electron but of the positron (*Positronensprung*). Third ($\mu = 3$), the positron changes its energy before the electron does. Fourth ($\mu = 4$), the positron changes its energy twice. In the fifth and sixth possibilities ($\mu = 5, 6$), two pairs are first created and the electrons and positrons are subsequently annihilated by crossing over, that is, the electron from the first pair and the positron from the second pair annihilate together and the positron from the first pair and the electron from the second pair (*Zerstrahlung ...*) annihilate together; the difference between the fifth and sixth possibilities is that either one or the other of the annihilations occurs first.

Each of these six possible sequences (*Reihenf.* = *Reihenfolge* in Fig. 2.1) has 24 possibilities as to which of the four photons is involved in the first partial process, which photon is involved in the second, which photon in the third and which in the fourth[7]:

---

[5] Heisenberg (1934).

[6] Euler (1936, p. 419).

[7] My translation of the original German text: "Hierbei kann die erste Paarerzeugung statt unter Absorption von g^1, wie in der Figur (und als Repräsentant in den folgenden Rechnungen)

Here, the first pair can also be created through the emission of $-g^3$, instead of through the absorption of g^1 as indicated in the figure (and as represented in the following calculations), etc. I. e. the 4 headings of the columns of the table in the first row (and in all subsequent formulae the 4 indices to the light quanta 1, 2, 3, 4), may be permuted if the remaining parts of the table (and formulae) are left unchanged.

Depending on the behaviour of the created pairs, there are 6 different transition paths denoted by $\mu = 1$ up to 6, and each of these 6 transition paths can be combined with all 24 permutations of the light quanta. (Euler 1936, p. 420)

The sum, in equation (2.2), over all the possibilities for the three intermediate electron–positron states k, e, m, can now be grouped as the sum: over the six possible sequences of creation, annihilation and energy change of the electrons and positrons; over the permutations of the involved photons; and as an integral over all the momenta that the initially created pair can have (which is not determined by energy conservation since we are examining virtual rather than real transitions here):

$$H^4_{\text{in}} = C \int d\mathfrak{p} \sum_{\text{Perm}} \sum_{\mu=1}^{6} \frac{Z_\mu}{N_\mu} + V^4_{\text{in}}, \tag{2.3}$$

Z_μ and N_μ being, up to constants, the numerator and denominator of (2.2) for each possibility μ. C is given by: the volume of the cube in which the electromagnetic field is supposed to be periodic; the momenta of the four light quanta involved; and some constant factors; $\mathfrak{p}$ is the momentum vector of the initially created electron–positron pair.[8]

Euler then goes on to evaluate the matrix element for two special cases in order to compare the results with the field theoretical expression for the interaction energy obtained in previous sections of his article. In this evaluation, Euler uses tables to list systematically all the contributions and their numerical values (see Fig. 2.2). The rows of the tables indicate the different permutations of the light quanta, while the columns indicate the term of the perturbative expansion that is evaluated in the cell of the table in the respective row and column.

Most studies on Feynman diagrams claim that the main achievement of Feynman diagrams was that tedious calculations, such as Euler's, could be avoided. For instance, David Kaiser (2005, pp. 35–37) finds that Euler's tables show that "by the mid-1930s QED seemed an unholy mess, as calculationally intractable as it was conceptually muddled". The solution to the problems, however, is, according to Kaiser,

angegeben, auch unter Emission von $-g^3$ vor sich gehen usw. D. h. es können die 4 Überschriften, die über den Spalten der Tabelle in der obersten Zeile stehen (und in allen folgenden Formeln die 4 Lichtquantenindizes 1, 2, 3, 4) noch in beliebiger Weise permutiert werden unter Beibehaltung aller übrigen Tabellen- (und Formel-) Teile.

Je nach dem Verhalten der erzeugten Paare gibt es 6 verschiedene Übergangswege, bezeichnet durch $\mu = 1$ bis 6 und jeder dieser 6 Übergangswege kann mit allen 24 Permutationen der Lichtquanten kombiniert werden."

[8] Euler (1936, pp. 420/421).

432 *Annalen der Physik. 5. Folge. Band 26. 1936*

Gl (8,10)

Reihenfolge	Zähler	1			$\frac{(\mathfrak{p}\,\mathfrak{g})}{p_0^2}$			$\frac{\|g\|^2}{p_0^2}$			$\frac{(\mathfrak{p}\,\mathfrak{g})^2}{p_0^4}$		
		1	$\frac{2p_y^2}{p_0^2}$	$\frac{8p_y^2p_z^2}{p_0^4}$	1	$\frac{2p_y^2}{p_0^2}$	$\frac{8p_y^2p_z^2}{p_0^4}$	1	$\frac{2p_y^2}{p_0^2}$	$\frac{8p_y^2p_z^2}{p_0^4}$	1	$\frac{2p_y^2}{p_0^2}$	$\frac{8p_y^2p_z^2}{p_0^4}$
$g^1g^2g^3g^4$	$z_1=$	0	0	−1	0	4	2	2	0	1	−2	−8	−4
	$z_2=z_3=$	0	0	−1	0	4	2	2	0	1	−2	−8	−4
	$z_4=$	0	4	−1	0	−4	2	2	−4	1	−2	8	−4
	$z_5=z_6=$	−8	8	−1	0	−8	2	4	−6	1	−4	14	−4
$g^1g^2g^4g^3$	$z_1=$	0	0	−1	0	0	0	−2	0	1	2	0	−2
	$z_2=z_3=$	0	0	−1	0	0	0	−2	0	1	2	0	−2
	$z_4=$	0	4	−1	0	4	0	2	−4	1	−2	8	−2
	$z_5=z_6=$	−8	8	−1	0	0	0	4	−6	1	−4	14	−2
$g^1g^3g^4g^2$	$z_1=$	0	4	−1	0	−12	4	2	−12	3	−2	40	−14
	$z_2=z_3=$	0	0	−1	0	0	4	−2	0	3	2	0	−14
	$z_4=$	0	0	−1	0	0	4	−2	0	3	2	0	−14
	$z_5=z_6=$	−8	8	−1	0	−16	4	4	−14	3	−4	46	−14
$g^1g^4g^3g^2$	$z_1=$	0	4	−1	0	−4	2	2	−4	1	−2	8	−4
	$z_2=z_3=$	0	0	−1	0	−4	2	2	−4	1	−2	8	−4
	$z_4=$	0	0	−1	0	−4	2	2	−4	1	−2	8	−4
	$z_5=z_6=$	−8	8	−1	0	−8	2	4	−6	1	−4	14	−4
$g^1g^4g^2g^3$	$z_1=$	0	0	−1	0	0	0	2	−4	1	−2	4	−2
	$z_2=z_3=$	0	4	−1	0	0	0	2	−4	1	−2	4	−2
	$z_4=$	0	0	−1	0	−4	0	−2	0	1	2	−4	−2
	$z_5=z_6=$	8	−8	−1	0	0	0	−4	2	1	4	−10	−2
$g^1g^3g^2g^4$	$z_1=$	0	0	−1	0	4	4	−2	4	3	2	−20	−14
	$z_2=z_3=$	0	4	−1	0	−8	4	2	−8	3	−2	20	−14
	z_4-	0	0	−1	0	0	4	2	−4	3	−2	4	−14
	$_5=z_6=$	8	−8	−1	0	16	4	−4	10	3	4	−42	−14

Für die *Reihenfolge*: $g^1g^4g^2g^3$ *und* $g^1g^3g^2g^4$

$$Z_\mu = -S_1 + 2p_y^2\left[-\frac{\lambda_\mu^3}{p_0^1\,p_0^3} + \frac{\lambda_\mu^2}{p_{03}\,p_0^4} + \frac{\lambda_\mu^2\lambda_\mu^3}{p_0^1\,p_0^4} - \frac{1}{p_0^2\,p_0^3} + \frac{2\lambda_\mu^2}{p_0^1\,p_0^3} - \frac{2\lambda_\mu^3}{p_0^2\,p_0^4}\right]$$
$$-\frac{1}{p_0^1p_0^2p_0^3p_0^4}\left[-S_2 + 2p_y^2\left[(\mathfrak{p}^1\mathfrak{p}^2)+(\mathfrak{p}^3\mathfrak{p}^4)+(\mathfrak{p}^2\mathfrak{p}^3)+(\mathfrak{p}^1\mathfrak{p}^4)-2(\mathfrak{p}^1\mathfrak{p}^3)-2(\mathfrak{p}^1\mathfrak{p}^4)\right]\right].$$

Darin bedeuten S_1, S_2 die Abkürzung:

$$S_1 = \lambda_\mu^2\lambda_\mu^3 - \lambda_\mu^3\frac{(\mathfrak{p}^1\mathfrak{p}^2)+mc^2}{p_0^1p_0^2} + \lambda_\mu^2\frac{(\mathfrak{p}^3\mathfrak{p}^4)+(mc)^2}{p_0^3p_0^4} - \lambda_\mu^3\frac{(\mathfrak{p}^2\mathfrak{p}^4)+(mc)^2}{p_0^2p_0^4}$$
$$+\lambda_\mu^2\lambda_\mu^3\frac{(\mathfrak{p}^1\mathfrak{p}^4)+(mc)^2}{p_0^1p_0^4} - 1\cdot\frac{(\mathfrak{p}^2\mathfrak{p}^3)+(mc)^2}{p_0^2p_0^3} + \lambda_\mu^2\frac{(\mathfrak{p}^1\mathfrak{p}^3)+(mc)^2}{p_0^1p_0^3},$$

H. Euler. Über die Streuung von Licht an Licht usw. 433

$\frac{(\mathfrak{p}\,\mathfrak{g})\,\|g\|^2}{p_0^4}$			$\frac{(\mathfrak{p}\,\mathfrak{g})^3}{p_0^6}$			$\frac{\|g\|^4}{p_0^4}$			$\frac{(\mathfrak{p}\,\mathfrak{g})^2\,\|g\|^2}{p_0^6}$			$\frac{(\mathfrak{p}\,\mathfrak{g})^4}{p_0^8}$		
1	$\frac{2p_y^2}{p_0^2}$	$\frac{8p_y^2p_z^2}{p_0^4}$	1	$\frac{2p_y^2}{p_0^2}$	$\frac{8p_y^2p_z^2}{p_0^4}$	1	$\frac{2p_y^2}{p_0^2}$	$\frac{8p_y^2p_z^2}{p_0^4}$	1	$\frac{2p_y^2}{p_0^2}$	$\frac{8p_y^2p_z^2}{p_0^4}$	1	$\frac{2p_y^2}{p_0^2}$	$\frac{8p_y^2p_z^2}{p_0^4}$
−4	−4	−4	4	16	8	−2	0	−1	10	16	12	−8	−32	−16
−4	−4	−4	4	16	8	−2	0	−1	10	16	12	−8	−32	−16
−4	12	−4	4	−16	8	−2	4	−1	10	−32	12	−8	+32	−16
−8	18	−4	+8	−26	8	$-\frac{7}{2}$	$\frac{11}{2}$	−1	19	−47	12	$-\frac{31}{2}$	$\frac{99}{2}$	−16
0	0	0	0	0	0	2	0	−1	−6	0	6	4	0	−6
0	0	0	0	0	0	2	0	−1	−6	0	6	4	0	−6
4	−6	0	−4	10	0	−2	4	−1	10	−24	6	−8	24	−6
0	0	0	0	0	0	$-\frac{7}{2}$	$\frac{11}{2}$	−1	19	−39	6	$-\frac{31}{2}$	$\frac{83}{2}$	−6
−12	78	−22	12	−138	48	−10	38	−9	66	−402	118	−56	488	−166
8	0	−22	−8	0	48	6	0	−9	−34	0	118	28	0	−166
8	0	−22	−8	0	48	6	0	−9	−34	0	118	28	0	−166
−16	84	−22	16	−148	48	$-\frac{23}{2}$	$\frac{79}{2}$	−9	75	−417	118	$-\frac{127}{2}$	$\frac{1011}{2}$	−166
−4	12	−4	4	−16	8	−2	4	−1	10	−32	12	−8	32	−16
−4	12	−4	4	−16	8	−2	4	−1	10	−32	12	−8	32	−16
−4	12	−4	4	−16	8	−2	4	−1	10	−32	12	−8	32	−16
−8	18	−4	8	−26	8	$-\frac{7}{2}$	$\frac{11}{2}$	−1	19	−47	12	$-\frac{31}{2}$	$\frac{99}{2}$	−16
0	0	0	0	0	0	−2	4	−1	6	−16	6	−4	12	−6
0	0	0	0	0	0	−2	4	−1	6	−16	6	−4	12	−6
−4	6	0	4	−10	0	2	0	−1	−10	8	6	8	−12	−6
0	0	0	0	0	0	$\frac{7}{2}$	$-\frac{3}{2}$	−1	−19	23	6	$\frac{31}{2}$	$-\frac{59}{2}$	−6
12	−38	−22	−12	82	48	10	−18	−9	−66	230	118	56	−316	−166
−8	40	−22	8	−56	48	−6	20	−9	34	−172	118	−28	172	−166
−8	24	−22	8	−24	48	−6	16	−9	34	−124	118	−28	108	−166
16	−60	−22	−16	124	48	$\frac{23}{2}$	$-\frac{47}{2}$	−9	−75	293	118	$\frac{127}{2}$	$-\frac{795}{2}$	−166

$$S_2 = \left[\begin{array}{l}\;\;[(\mathfrak{p}^1\mathfrak{p}^2)+(mc)^2]\,[(\mathfrak{p}^3\mathfrak{p}^4)+(mc)^2] + [(\mathfrak{p}^2\mathfrak{p}^3)+(mc)^2]\,[\mathfrak{p}^1\mathfrak{p}^4)+(mc)^2]\\ -\{(\mathfrak{p}^1\mathfrak{p}^3)+(mc)^2]\,[(\mathfrak{p}^2\mathfrak{p}^4)+(mc)^2]\end{array}\right].$$

Diese Ausdrücke werden bei weiterer Spezialisierung auf parallele Impulse, wie man durch Einsetzen von (8,8, δ) und Entwicklung nach g/mc bis zur 4. Ordnung findet:

Vgl. Gl. (8,10).

Um die Zähler (5,15) durch die Nenner (5,14) zu dividieren, über die 6 Übergangswege μ und die 24 Reihenfolgen (8,8, α) zu summieren, muß man die Spalten der Tab. (8,10) mit den entsprechenden Spalten der Tab. (8,9) und mit dem Faktor 4 multiplizieren. Man erhält dann im Glied 4. Ordnung in g/mc nach Mittelung über die Positronenrichtung $\mathfrak{p}$: [vgl. § 9,7]

Annalen der Physik. 5. Folge. 26. 29

Fig. 2.2 Euler's evaluation of a special case of light-by-light scattering: "Equation (8.10)", which extends over two pages (Euler 1936, pp. 432–433)

not to be found in a "sweeping conceptual revolution" or "great epistemological upheavals" but by "focusing on the evaluation of specific integrals".

Similar views are also shared in the first publications to use Feynman diagrams. In one of the first treatments of light-by-light scattering using renormalized QED and Feynman diagrams (see Fig. 2.3), Robert Karplus and Maurice Neuman Karplus (1950, p. 380) emphasize that the new theory of QED provides researchers with more powerful calculational tools:

> Since the corrections we are discussing are necessarily at least of the order e^4, their calculation has involved considerable complications both because the treatment of effects involving virtual pairs has been traditionally accompanied by divergence and gauge-invariance difficulties and because the expressions encountered were lengthy and tedious to manipulate. With the promise the recent developments in quantum electrodynamics give of eliminating the former and reducing the latter of these obstacles, it seemed worth while to re-examine the problem in spite of the smallness of the effects and the consequent difficulties attending their experimental detection.

It is undoubtedly true that Feynman diagrams help to sort out the various terms in a perturbative calculation. They certainly make it much easier than Euler's treatment to recognize when terms are identical or zero. Nevertheless, it can still be hard to evaluate the terms corresponding to a Feynman diagram, and long lists of formulae and tables of various terms are not unusual in such evaluations.[9]

Rather than allowing one to do away with the need to carry out long calculations using tables, Feynman diagrams change the way in which light-by-light scattering is explained. Feynman diagrams are the successors to Euler's versions of *term schemes*, and not the successors to Euler's *tables* (as Kaiser's account suggests). Having replaced Euler's term schemes with Feynman diagrams is evidence that the theoretical model for the phenomena, for which one was calculating cross sections and other characteristic quantities, has changed. Only to a lesser extent do Feynman diagrams serve the same function as the systematic lists of perturbative terms; they serve mainly to represent a new model.

The new concept of light-by-light scattering as a particular sequence of particle creation, propagation and annihilation described by field theoretical vacuum expec-

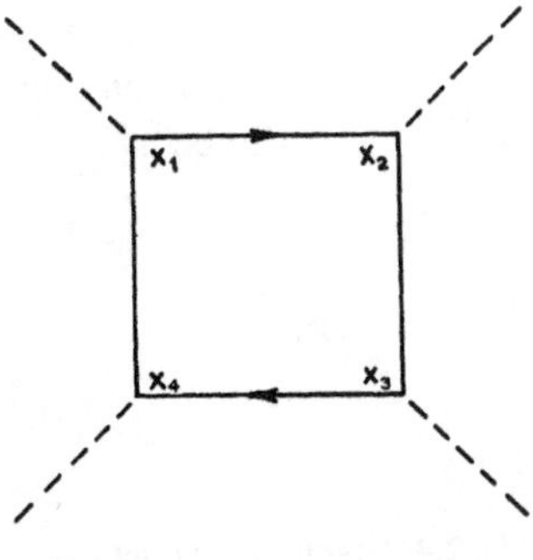

Fig. 2.3 A Feynman diagram used in the treatment of light-by-light scattering (Karplus and Neuman 1950, Figure 1)

[9] See, for instance, Devoto and Duke (1984).

tation values is an innovative result and was clearly not invented with the primary goal of facilitating calculations. It is not a model that researchers had in mind, and only Feynman and Dyson find a set of rules to translate the graphical representation of this supposedly well-known model into the correct field theoretical expressions. Even then, Feynman, in a lecture he gives at Cornell University (Ithaca, NY) around 1948, mistakenly draws the wrong diagram to represent light-by-light scattering (see Figs. 2.4 and 2.5, see also Kaiser 2005, p. 187).

Diagrams with the same kind of syntactic elements as Feynman diagrams are not used to illustrate which intermediate processes might bring about the scattering of light by light but only to display the *kinematical* relations of the studied phenomena (see Fig. 2.6; cf. Fig. 2.7).

Another case in point is Compton scattering, the Feynman diagrams of which are shown in Chapter 6 (see Fig. 6.6). Since Arthur Compton (1923) the interaction of light quanta off electrons had been described as a scattering effect. Before, it had not been entirely clear whether the light quantum could be treated as an ordinary particle with respect to its kinematical properties such as energy and momentum.

The typical graphical representation of the *kinematical* characteristics of a situation such as that of the scattering of an electron and a photon displays the direction of the momenta of the particles (see Fig. 2.7). However, when it comes to the description and theoretical treatment of such kinematical observations as certain processes that bring these about, scattering diagrams are not used. Rather, the situation is conceived of as a sequence of transitions between the energy levels of the electron caused by the absorption or the emission of a light quantum (see Fig. 2.8).

Looking back, it is hard to conceive that representations similar to Feynman diagrams were not being used by scientists working at the time to represent scattering situations, particularly as, in today's context, readers of textbooks seem tempted to take Feynman diagrams to represent the scattering of an elementary particle in an almost classical way (see Section 1.2). How you can then connect this representation to the correct quantum field theoretical description remains mysterious. It would seem that only someone as bold as Feynman would have dared to apply this pre-quantum theoretical representation to quantum theoretical problems, and be rewarded for his boldness with finding a one-to-one correspondence to field theoretical expressions. Correct results could, it seems, be derived by using the diagrams only as mnemonic devices. That the diagrams were misleading (because they could easily be taken to represent quantum field theoretical scattering processes

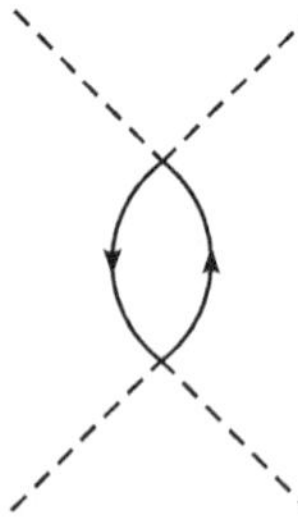

Fig. 2.4 Feynman's wrong Feynman diagram! Detail of Fig. 2.5, adapted by the author (A. W.)

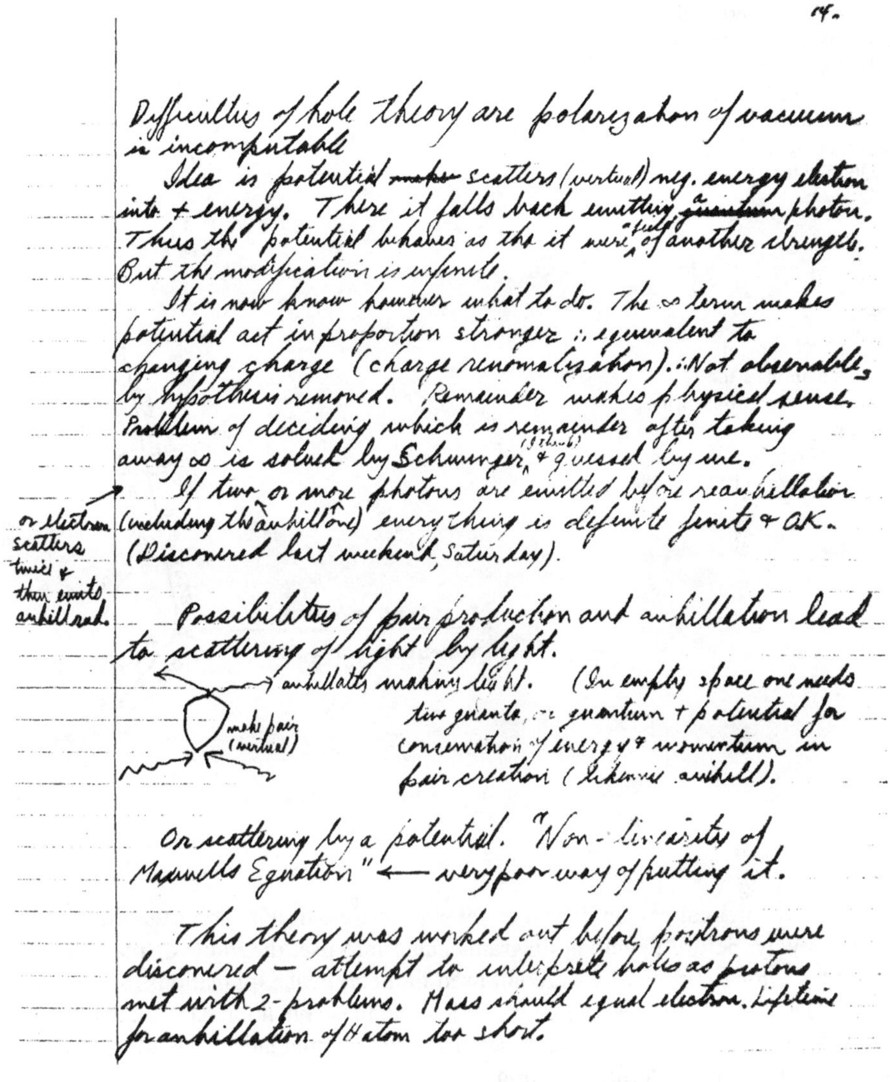

14.

Difficulties of hole theory are polarization of vacuum is incomputable

Idea is potential ~~makes~~ scatters (virtual) neg. energy electron into + energy. There it falls back emitting a ~~quantum~~ photon. Thus the potential behaves as tho it were itself of another strength. But the modification is infinite.

It is now know however what to do. The ∞ term makes potential act in proportion stronger ∴ equivalent to changing charge (charge renormalization). ∴ Not observable, by hypothesis removed. Remainder makes physical sense. Problem of deciding which is remainder after taking away ∞ is solved by Schwinger, (I think) & guessed by me.

If two or more photons are emitted before reannihilation (including the annihilation) everything is definite finite & OK. (Discovered last weekend, Saturday).

or electron scatters twice & then emits annihilated.

Possibilities of pair production and annihilation lead to scattering of light by light.

annihilates making light. make pair (virtual)

(In empty space one needs two quanta, or quantum + potential for conservation of energy & momentum in pair creation (likewise annihil).)

On scattering by a potential. "Non-linearity of Maxwells Equation" ← very poor way of putting it.

This theory was worked out before positrons were discovered — attempt to interpret holes as protons met with 2-problems. Mass should equal electron. Lifetime for annihilation of H atom too short.

Fig. 2.5 Even Feynman drew the wrong Feynman diagrams!: a diagram in his lecture notes on light-by-light scattering, ca. 1948 (*Advanced Quantum Mechanics (at Cornell)*, folio 5, reprinted with permission of Melanie Jackson Agency, LLC)

as processes of classical mechanics) seems to be the price to pay for their usefulness in carrying out complex calculations.

However, it needs to be made clear that Feynman diagrams (certainly by the time that Dyson systematized the diagrams) are not classical representations of scattering events. They represent the latter's relevant aspects in a more abstract way: they show the connections between the creation and annihilation events, and these connections are understood to be the vacuum expectation values of field operators. This highly

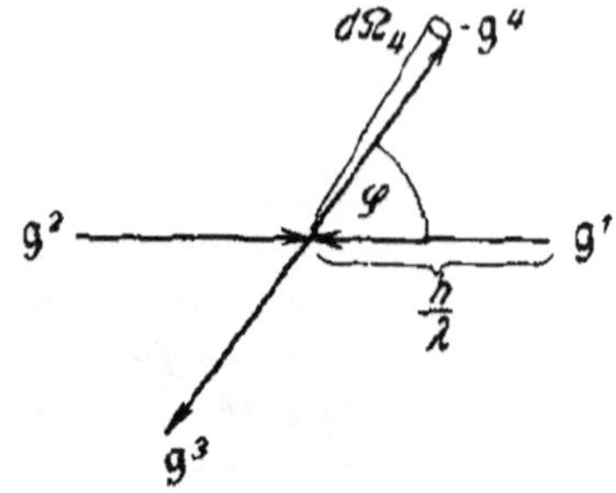

Fig. 2.6 A scattering diagram showing the dynamical relationships in the calculation of the light-by-light scattering cross section (Figure 3 in Euler 1936, p. 446)

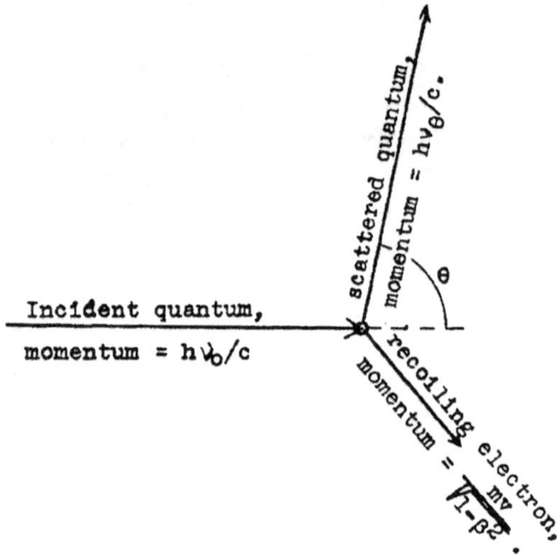

Fig. 2.7 Graphical representation of the kinematical relations of the scattering of an electron and a photon (Compton scattering). From Compton (1923, p. 486)

elaborate theoretical content was unknown at the time of the earlier theories, which explains why, before Feynman diagrams came into use, diagrams displaying scattering trajectories or the like were not used in theoretical explanations of scattering events. The internal processes that brought about these phenomena were understood to be the transitions between energy levels; only the kinematical relations classified the phenomena as scattering situations.

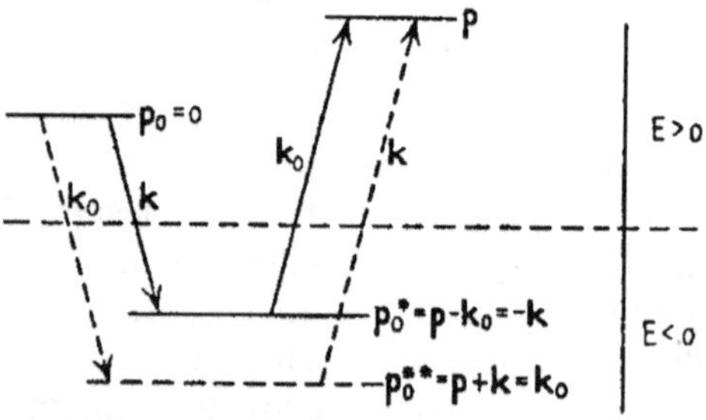

Fig. 2.8 Diagrammatic representation of the scattering of an electron and a photon by the adaptation of term schemes (from the standard textbook by Heitler (1944, p. 190); this drawing was withdrawn from the later third edition (1954)

2.2 Classifying Possible Transitions

Because of the Second World War, physicists working in Japan were isolated from European and American physicists for some time and were only vaguely aware of theoretical developments in the West. Within the Japanese community of physicists, Euler's diagrams survived long enough to be used in attempts to eliminate the uninterpretable infinities of QED. Koba and Takeda adopted and modified them in order to classify the great many possible intermediate states, from an initial state to a final state, through which certain processes can pass. The classification allows them to show that the renormalization of the mass and charge of the electron is sufficient to make the infinities cancel each other out.

Their method has much in common with what would eventually be Dyson's generally accepted procedure. The reasons why Dyson's and not their method became the accepted procedure may be sociological or even political in nature. However, I will point out some of the shortcomings of the Japanese transition diagrams that might equally well explain the preference of the scientific community for Dyson's method. The Japanese transition diagrams were still too tightly rooted in the tradition of representing the energy levels of material systems in order to be able to articulate the relevant aspects of quantum electrodynamic phenomena.

In a series of three papers, Koba and Takeda (the first paper co-authored with Sin-Itiro Tomonaga, not Takeda) calculate the radiative corrections to certain types of scattering situations. They propose a "self-consistent subtraction method", which is essentially the same idea as renormalization. The divergences that occur in certain terms of the perturbation expansion are incorporated into a redefinition of the mass and charge parameter of the theory. A basic feature of their method is, accordingly, to classify all the terms of the perturbative expansion into corrections of "mass type" or "polarization type".[10]

In the first paper,[11] Koba and Tomonaga begin by correcting the elastic scattering of an electron in a potential (produced, for instance, by the electromagnetic field of another electron). Their goal is to modify the Hamiltonian, describing the interaction in such a way that the divergent terms cancel each other out. The divergent terms occur in the expression for the self-energy of the electron and in the expression for the polarization of the vacuum.

Koba and Tomonaga (RadReacI, p. 291) refer to Victor Weisskopf (1939), who calculated the self-energy of the electron—manifesting itself in a correction, δm, to the theoretical mass parameter, m, of the electron—as

$$\delta m = \frac{3e^2 m}{2\pi} \int_{[k]\neq 0}^{\infty} \frac{dk}{k} + \text{finite terms}, \tag{2.4}$$

---

[10] RadReacI; RadReacII; RadReacIIIa; RadReacIIIb. See also Sections 2.4.1 and 2.4.2.

[11] RadReacI.

where k are the momenta, or the "wave numbers" (Weisskopf 1939, p. 81), of the electromagnetic field (they set $\hbar = c = 1$).

"In accordance with Bethe's [Bethe 1947] assumptions", that is, assuming small (non-relativistic) velocities and energies, Koba and Tomonaga (RadReacI, p. 291) incorporate this infinite quantity into the definition of the observable mass of the electron. That is, the observable mass $\dot{m}$ is the difference of two infinite contributions: the original mass parameter in the Hamiltonian m and the self-energy δm, that is[12]

$$\dot{m} = m - \delta m. \tag{2.5}$$

Then they modify the Hamiltonian describing the *free* electron and the free electromagnetic fields by replacing m with the mass parameter $\dot{m}$. They choose to maintain the *total* Hamiltonian unmodified (p. 292); therefore, the *interaction* Hamiltonian must be modified in order to compensate for the change in the free Hamiltonian. This new interaction term gives rise to two new possibilities for intermediate transitions, or "connections" (p. 294), which the authors represent graphically (see Fig. 2.9).

The perturbative expansion also contains an infinite term that describes the polarization of the vacuum. The authors apply the same method to this type of infinity that they applied to removing the infinities related to the mass of the electron. Other methods of treating this type of infinity seem ad hoc to them:

> Let us consider the right-hand side of (1·5) [a formula similar to (2.4)] which has been so far *ad hoc* struck off as the vacuum polarization effect caused by the scattering potential. We now try to apply our self-consistent subtraction method also to this type of divergence, then we find that by this new method the same result—at least in the approximation under consideration—is obtained, but with a more plausible reasoning for the subtraction procedure. (RadReacI, p. 297, see also p. 292)

The infinite term is interpreted as arising from two processes, which are again represented by a graphical representation that displays the transitions from the initial state to possible intermediate states and then to the final state (see Fig. 2.10).

The authors determine the interaction Hamiltonians that describe these two processes. Incorporating these interaction Hamiltonians into the original Hamiltonian is tantamount to replacing the ideal potential V with the *observed* potential $\dot{V} = V + \delta V$.

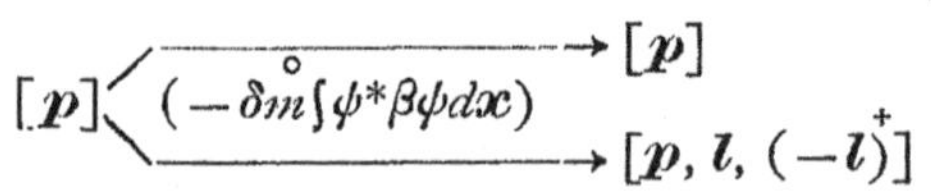

Fig. 2.9 Two new transitions to intermediate states induced by a change in the interaction Hamiltonian to compensate for the change in the free Hamiltonian of replacing the "mechanical" mass m with the observable mass $\dot{m}$ (RadReacI, p. 294; cf. Figure 4.4 (top) in Kaiser 2005, p. 137)

---

[12] The original has a plus sign rather than a minus sign; see also page 293 of the paper. As far as I can see, it should be a minus sign. Bethe (1947, p. 340) also says that one should "subtract".

$$\text{i)}\quad [p] \xrightarrow{\text{V}} [p, l, (-l-p+q)^{+}] \xrightarrow{\text{Coulomb}} [q]$$

$$\text{ii)}\quad [p] \xrightarrow{\text{Coulomb}} [q, l, (-l+p-q)^{+}] \xrightarrow{\text{V}} [q]$$

Fig. 2.10 The processes that cause the infinite polarization of the vacuum (RadReacI, p. 297).

In this first paper, the graphical representations, which display, by means of arrows, the possible transitions from the initial to intermediate states (and to an eventual final state) serve to identify a particular process. In the first case, the diagram shows the new processes that are possible due to modifying the interaction Hamiltonian by incorporating the self-energy of the electron. In the second case, the diagrams show the two processes responsible for the infinite vacuum polarization. Diagrammatic representations similar to these will become more important in their later papers in which they consider more complex initial states and higher-order corrections.

The authors do not claim to have resolved the difficulties of interpreting the divergences that arise in the standard formulation of QED. But they are able to classify the divergences, which allows them to subtract the infinities justifiably:

> It is true that our method does by no means give the real solution of the fundamental difficulty of the quantum electrodynamics, but it has, at least, revealed the nature of various diverging terms appearing in our scattering problem and reduced them into two quantities: the self-energy and the vacuum polarization. In this way it becomes possible in an unambiguous and consistent manner to treat the field reaction problem without touching the fundamental difficulty by employing the finite empirical values instead of the infinite "theoretical" values for these two quantities. (RadReacI, pp. 301–302)

In the second paper of the series (RadReacII), Koba and Takeda treat the next more complex initial state: instead of one electron that is scattered by an external potential, they now consider Compton scattering, that is, the scattering of an electron by a light quantum. In this article, similar schemes as in the first are used, albeit more intensively. Now they do not just identify the processes under discussion; they try above all to "classify" the correction terms in order to find a justifiable way of subtracting the infinite ones among them. They also consider higher-order corrections, which makes the mathematical description significantly more complicated. For the fourth-order corrections there are 1,080 "connections" to be considered from the initial to the final states, via intermediate states. A classification of all these possibilities into classes that will facilitate the elimination of the divergences is carried out by graphical representations, as shown in Fig. 2.11. The authors can easily write down the mathematical expression for the correction terms arising from the processes systematically listed in these "connection" schemes. The resulting expressions allow the authors to modify the interaction Hamiltonian in order to eliminate the divergences:

> Now that we have investigated the diverging terms thoroughly and have arranged them in the forms of [the mathematical expressions obtained after having listed the processes by connection schemes], it is not difficult to seek for a new interaction term to be introduced in order to eliminate infinite terms. (RadReacII, p. 413)

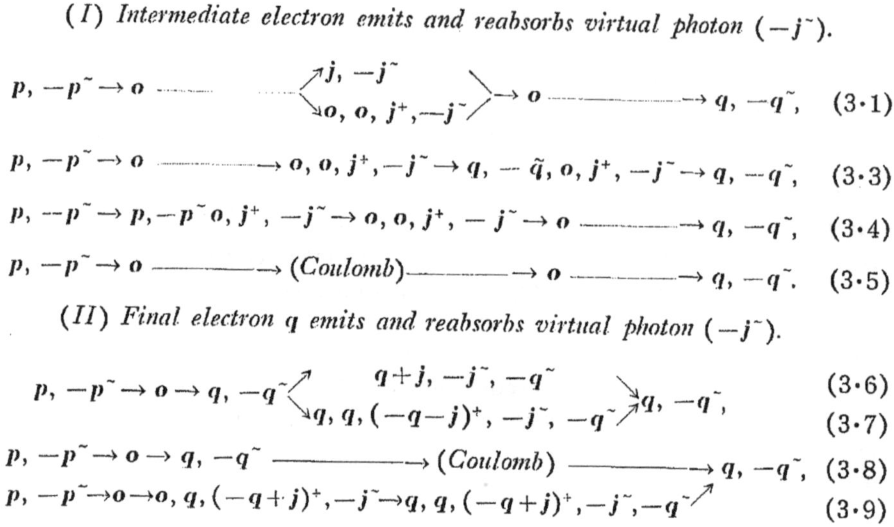

Fig. 2.11 The graphical representation used to classify radiative corrections to Compton scattering. Without giving any reasons, the authors warn the reader that the "titles should not be taken too literally" (RadReacII, p. 411)

2.3 An Attempt to Improve the Representation

After having treated elastic electron–electron scattering (Møller scattering) and electron–photon scattering (Compton scattering) in the first two papers of their series on "radiation reaction in collision processes", in their third paper (which is divided into two parts) Koba and Takeda proceed to the general case. For the complexities of the general case, the schemes used in the previous papers are no longer sufficient to classify the possible sequences of initial, intermediate and final states:

> In order to dispose of all possible radiative corrections in a general case, it is essential to analyze systematically the complicated chains connecting the initial and the final states not only through the least necessary number of intermediate states, but also through certain detours including the emission and reabsorption of a virtual photon or the creation and annihilation of a virtual electron–positron pair, which just account for the radiation reaction. For this purpose we have introduced a "transition diagram method," which turns out an effective tool for the discussion of higher order processes. The first half of the paper is appropriated to the illustration of this method, as our later reasoning is based entirely upon it. (RadReacIIIa, p. 61)

For a more appropriate representation, they find, among others, Euler's diagrams employed in light-by-light scattering. They take up the main features of these hole theoretical term schemes and adapt them for their purposes of identifying, interpreting and eliminating the divergences of quantum electrodynamic scattering processes.

In order to command a view of a whole connection between the initial and the final states which appears in the perturbation calculus of a certain complicated process, we propose here an improved form upon its diagram expression heretofore in use [reference to Euler (1936), among others]. (RadReacIIIa, p. 61)

As stated in the first citation of this section, the authors devote much of the third article to introducing their diagrams, and they base their arguments on this representation of the phenomena. First they introduce "notation and rules". "Notation" refers to the system of symbols used to explain which syntactical elements of the diagrams represent which parts of the physical system. A transition diagram consists of a connected sequence of line segments in a coordinate system of momentum and time. Electrons and positrons are represented as horizontal line segments. The presence or absence of photons is not explicitly represented but has to be inferred from the order of emission or absorption events represented by vertical lines:

An electron with momentum p is represented as a horizontal line segment at the height p with an arrow-head that points to the right [Fig. 2.12a] while a positron with momentum p—we denote it by p^+—is represented again by a horizontal segment, but at the height $(-p)$ and with an arrow-head directed to the left [Fig. 2.12b]. By the "height" in these figures is meant, of course, a three-dimensional position in the momentum space. The meaning of the arrow will be clarified later.

The existence of photons, real or virtual, need not be shown explicitly; only their emission or absorption is indicated by the "leaps", i. e. vertical segments connecting the two horizontal lines which correspond to the two momenta of electron or positron concerned. The emission we denote by a ripple-line and the absorption by a dotted line. Here also an arrow-head is affixed to indicate the direction, so that we may distinguish a photon with momentum k—we denote it by $\tilde{k}$—from $(-\tilde{k})$. (RadReacIIIa, pp. 61–62)

In a moment we will see that this inability to show the existence of photons is one of the shortcomings of the Japanese transition diagrams. For, in order to reflect certain relevant differences between the processes, one needs to be able to articulate explicitly the presence or absence of photons as well as the process of their creation or annihilation.

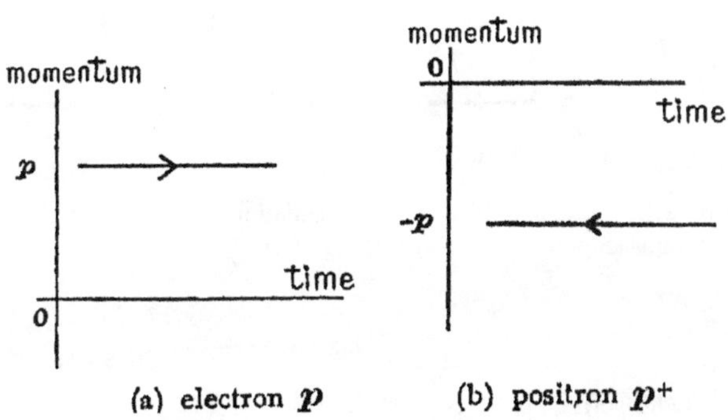

Fig. 2.12 Representation of an electron and a positron with momentum p (Figure 1 in RadReacIIIa, p. 62; cf. Figure 4.5, Kaiser 2005, p. 136)

Using this notation the authors can then represent eight "fundamental processes", which are described by the first-order term of the perturbative expansion. The authors give the following list[13]:

(a) an electron p emits a photon $\tilde{k}$ and goes over to $p - k$;
(b) a positron p^+ emits a photon $\tilde{k}$ and goes over to $(p - k)^+$;
(c) an electron p absorbs a photon $\tilde{k}$ and goes over to $p + k$;
(d) a positron p^+ absorbs a photon $\tilde{k}$ and goes over to $(p + k)^+$;
(e) an electron–positron pair p, q^+ is created with the emission of a photon $(\widetilde{-p-q})$;
(f) an electron–positron pair p, q^+ is annihilated with the emission of a photon $(\widetilde{p+q})$;
(g) an electron–positron pair p, q^+ is created with the absorption of a photon $(\widetilde{p+q})$;
(h) an electron–positron pair p, q^+ is annihilated with the absorption of a photon $(\widetilde{-p-q})$.

The graphical representation of these eight fundamental processes is shown in Fig. 2.13.

Only diagrams in which the line segments are composed such that the direction of the arrows is always the same when following the line from one end to the other are permitted. By following this rule only processes that satisfy momentum conservation can be represented. It is unnecessary to take energy conservation into account, since in transitions to intermediate states energy conservation may be violated:

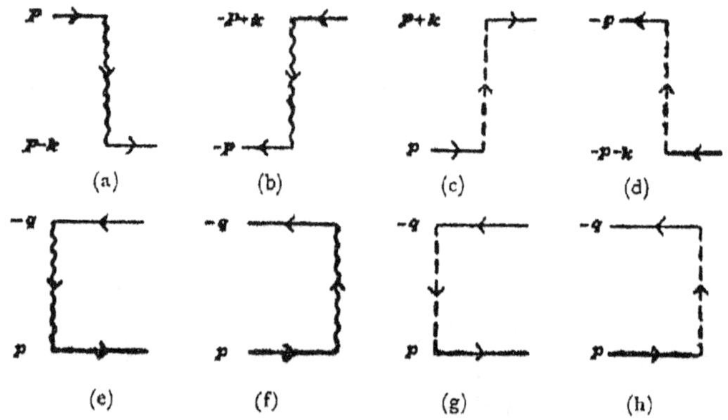

Fig. 2.13 Koba and Takeda's representation of the eight fundamental processes using transition diagrams (Figure 3 in RadReacIIIa, p. 63)

13 RadReacIIIa, pp. 63–64

> One can see from [Fig. 2.13] at once that, with the conventions made in the preceding section, the condition of momentum conservation is automatically fulfilled if one connects the three segments, which represent a light quantum and two particles concerned, in a proper order, i. e. in such a way that the arrows always follow one another. (RadReacIIIa, p. 64)

The "notation" interprets the graphical elements; the "rules" define what kind of *complex* diagrams have a meaning. For instance, it would be impossible to ascribe a meaning to the diagrams in which the flow of the arrow is inverted at least once; this would correspond to a process violating momentum conservation and is, therefore, not permitted.

All these explications of "rules" and "notation" are a preparation for an argument to come:

> It will be convenient for the later argument to call the arrow direction in the electron- or positron-line the direction of the "proper time" of that particle—in contrast to the "field time" t (abscissa), which flow always from left to right, and according to which the sequence of intermediate states should be traced. (RadReacIIIa, p. 64)

This arrow convention is similar to the one used in Feynman diagrams and is linked to the possibility of understanding the positron as an electron going backwards in time:

> In this sence [sic] one might be allowed to regard a positron metaphorically as "an electron whose proper time flows in the opposite direction to the field time", and the pair creation or annihilation as a sudden inversion of the direction of the proper time of one (not two!) particle [sic]. (RadReacIIIa, p. 64)

This concept of the positron is reminiscent of, indeed is almost identical to, Feynman's concept in his "Theory of Positrons", (see, e. g., p. 753), and in "Space-Time Approach to Quantum Electrodynamics", see, e. g., fn. 12), and to the ideas of E. C. G. Stueckelberg.[14] However, the Japanese authors cite a certain G. Zisman[15] and Marïo Schönberg[16] as originators of the idea. They knew very little of Feynman's work, as they explicitly state in an addendum at the end of their article.[17]

One of the examples the authors give for representing a scattering situation in their transition diagrams is Compton scattering (see Fig. 2.14; cf. Figs. 2.7, 2.8 and 2.11).

The transition diagram of a certain process can also be translated into a "representation by sequences". The two sequences of a transition diagram are obtained as follows. For the first sequence follow the lines of the diagram in the direction indicated by the arrows and note, in order of appearance, when a photon is absorbed

---

[14] Stueckelberg (1941).

[15] The authors give the following reference: "G. Zisman, Journ. Exp. Theor. Phys. (in Russian) 10 (1940), 1163." This is an article in the Russian journal "Zhurnal eksperimental'noi i teoreticheskoi fiziki", which, between 1955 and 1992, is translated and published by the American Institute of Physics as *Soviet Physics–JETP* and afterwards continued under the title *Journal of experimental and theoretical physics*.

[16] Schönberg (1946).

[17] RadReacIIIa, p. 141.

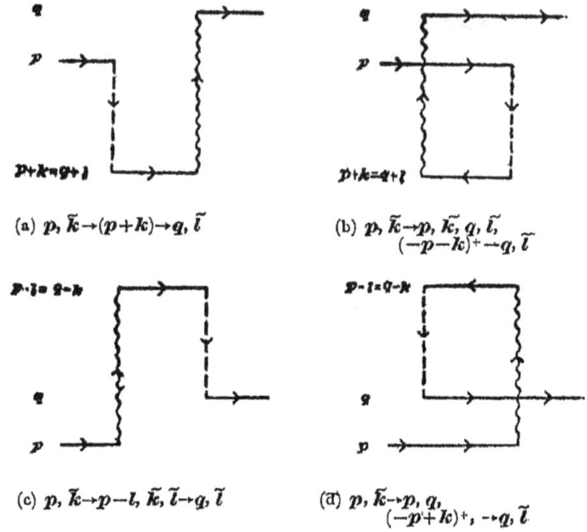

Fig. 2.14 Koba and Takeda's representation of Compton scattering using transition diagrams (Figure 4 in RadReacIIIa, p. 64; cf. Figure 4.6 in Kaiser 2005, p. 136)

or emitted by writing down its momentum, marked with an asterisk in the case of emission. The order of emission and absorption events so obtained is relative to proper time. The second characteristic sequence is obtained by the emission and absorption events according to field time. That is, one has to work through the diagram from left to right and note, as before, whether a photon is emitted or absorbed and with what momentum. For the transition diagram in Fig. 2.14 d, for instance, the characteristic pair of sequences is $(l^*, k|k, l^*)$: if one follows the line of the diagram in the direction indicated by the arrows, the first vertical segment one encounters is the wiggly line representing the emission of a photon with momentum l, which changes the electron's momentum from its initial value p to $p - l$. Therefore, the first symbol we write down is "l^*". The second vertical element one encounters is the broken line representing the absorption of a photon with momentum k, which gives the electron its final momentum $q = p - l + k$; we write down "k". If we work through the diagram from left to right, that is, if we follow the field time instead of the proper time, we come across the vertical elements in the inverse order, such that the second characteristic sequence is $k,\ l^*$.

The representation by transition diagrams and the corresponding sequences of the processes allow the authors to calculate the number of different ways in which a certain transition, from a given initial to a given final state, can occur.

> With the help of our diagram expression, one can classify any irreducible process including electrons, positrons, and photons by two numbers N and L. By an irreducible process is meant one that cannot be reduced into two or more simpler processes occurring at the same time but independently one upon another. (RadReacIIIa, pp. 65–66)

Every transition (for example, Compton scattering) can be classified according to the number of sets of connected line segments N and number of leaps L. Given N and L, the number of possible transitions is determined. For the case of Compton scattering, for instance, we have $N = 1$ and $L = 2$. The characteristic sequences have, therefore, two elements each; and there is only one line to follow ($N = 1$). The number of possible transitions is determined by the number of possible pairs of sequences. In the case of Compton scattering, we have two possibilities for each sequence, giving four possibilities for pairs of sequences (therefore, also four possible transition diagrams; see Fig. 2.4). In fact, the authors start with the slightly more general case where $N = 1$ and L is arbitrary:

> For the given value of N, L [or total number of real photons in initial and final states Q], one can now calculate the number of possible connections. First we treat the simple case $N = 1$, $L(= Q)$ =arbitrary. Here the diagram consists of only one set of segments as in [Fig. 2.4] and we have seen [in the previous section] that to such a possible connection always corresponds a pair of sequences each containing L elements. The converse is also true, as can be verified by drawing a scheme actually. The one-to-one correspondence being verified, the number of possible connections is nothing but that of possible ways of forming two sequences out of given L elements $(L!)^2$. (RadReacIIIa, p. 66)

By using similar combinatorial considerations concerning the possible compositions of graphical elements, the authors can calculate the number of possible processes for more general cases. Throughout this introduction to their diagrammatic method of classifying transition processes, the authors restrict themselves to the lowest order that describes the transition in the perturbative expansion. These terms, however, are unproblematic; they do not show any divergences. The latter only appear in the next-to-leading term, in the so-called radiative corrections.

The authors provide graphical rules for constructing a new transition diagram that corresponds to the first radiative correction to a given leading-order transition from the old transition diagram. The diagram representing the correction is obtained by adding two new leaps: an absorption and an emission of a virtual photon with arbitrary momentum j. The graphical constraints show that there are two first radiative corrections:

> Suppose that a certain process with its all possible connections is given. For simplicity we assume $N = 1$, $L = Q$. In order to calculate the first radiative correction, we have to construct connections with $L + 2$ elements, that is, the given Q and j^* and j out of the given connections with L elements. This can be performed in two ways: a) one can add the two new elements j^* and j to the existing line, so that the latter receives a swelling-like correction [Fig. 2.15, left]; b) one can build apart from the existing one a new closed line including j^* and j, thus resulting in an island-like correction [Fig. 2.15, right]. It will be seen in the following that the first kind correction is what we have called of the

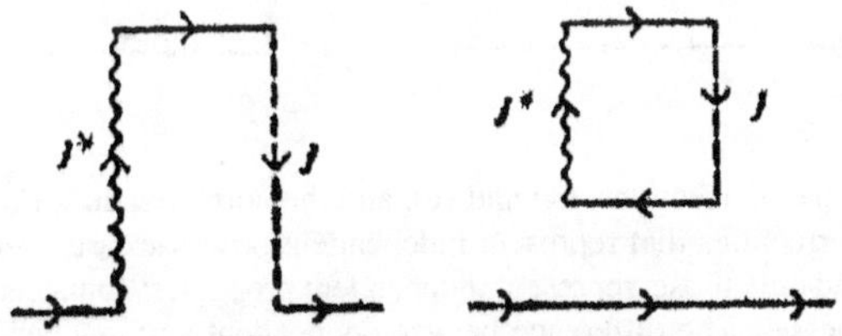

Fig. 2.15 A "swelling-like correction" (*left*) and an "island-like" correction (*right*). Square well-like corrections, and only those, are of the "mass-type"; island-like corrections, and only those, are of the "polarization-type" (Figure 6 in RadReacIIIa, p. 68)

mass-type $[M]$, and the second what we have called of the polarization-type $[P]$ in [RadReacII]. (RadReacIIIa, p. 68)

Although the transition diagrams appropriately articulate the distinction between types of mass and polarization corrections, they do not do so for the relevant differences among the polarization-type corrections, not all of which have to be taken into account. The only relevant polarization corrections are those in which a photon of the uncorrected process is involved:

> Out of the polarization type corrections the most general one is, using the notations of [RadReacII], the process
>
> $$\cdot \rightarrow r, (-r - j)^{+}, \tilde{j} \rightarrow \cdot \tag{2.6}$$
>
> where both r and $\tilde{j}$ are quite arbitrary. Such a process, however, cannot in general affect the cross section at all, because similar contributions, which come from various distribution [sic] in field time, when summed, vanish on account of interference with the original process, a result to be expected for any two independent events. (RadReacIIIa, p. 68/69)

The problem with these transition diagrams is that the independence of the graphical elements within them (the "island" is independent of the line) does not correspond one-to-one to the physical independence of the processes. Only some, not all, of the processes represented by an independent island are physically independent of the other partial processes. And this difference is relevant, since the physically independent processes do not have an observable effect and do not have to be taken into account when calculating observable quantities.

Feynman diagrams correctly articulate the dependence or independence of such polarization processes (see Fig. 2.16). In RadTh, Dyson will prove that the disconnected parts of Feynman diagrams can be ignored in the calculations (see Section 6.4.1). However, he immediately comments on his proof by saying that such proof is hardly needed:

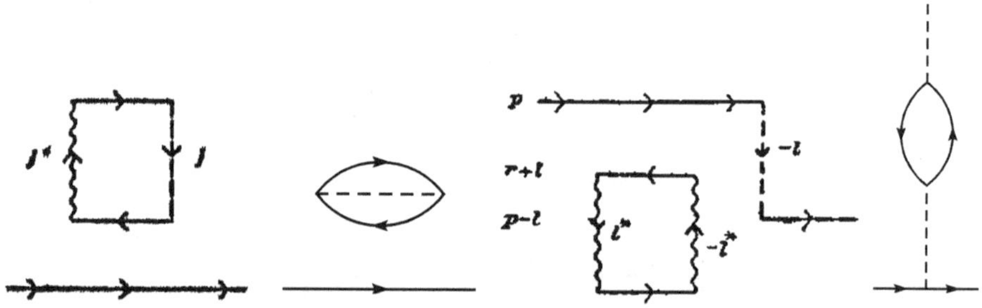

(a) (b) (c) (d)

Fig. 2.16 Two island-like corrections, (**a**) and (**c**), and the corresponding Feynman diagrams, (**b**) and (**d**). Island-like corrections that represent independent processes are irrelevant. Disconnected Feynman diagrams, and only those, represent independent processes; connected diagrams, and only those, dependent processes. The difference between dependent and independent processes is not articulated in Koba and Takeda's momentum diagrams (details of Figures 6 and 7 in RadReacIIIa, pp. 68/69; Feynman diagrams by A. W.)

> The elimination of disconnected graphs is, from a physical point of view, somewhat trivial, since these graphs arise merely from the fact that meaningful physical processes proceed simultaneously with totally irrelevant fluctuations of the vacuum. (RadTh, p. 496)

On a graphical level Feynman diagrams reflect these physical reasons in such an appropriate way that, when using them, the irrelevant quantities are almost automatically left out.

2.4 Cancellation of the Infinities

After having explained how to construct the transition diagrams for the first radiative correction, Koba and Takeda can, using the same graphical and combinatorial methods as with the leading-order processes, calculate the number of correction terms. It is important to know how many there are when it comes to avoiding leaving any of them out: the systematic elimination of the divergences hinges on subtly accounting for which terms cancel each other out.

The transition diagrams, however, do not only help one to calculate combinatorially the number of radiative corrections. They also help to classify the radiative correction terms in groups that cancel each other out and to recognize when some terms are equal. This is particularly important when the radiative corrections are divergent, as is mostly the case. The diagrams even help one to find the terms in the first place.

2.4.1 Mass-Type Divergences

Not all correction terms are divergent (RadReacIIIa, p. 70). In the second half of their third paper (RadReacIIIb), Koba and Takeda identify the divergent terms and show how they can be eliminated. The authors first treat the "mass type" divergences. They give a synopsis of the "connections" that could possibly yield divergent terms (see Fig. 2.17). Figure 2.18 shows the transition diagrams corresponding to the first group of sequences.

This representation of the processes identifies them as describing the self-energy of the electron in the intermediate state, since the diagrams are, apart from a horizontal reflection, identical to the diagrams that represent the self-energy of an electron in general—a "simple example" of the application of their transition diagrams considered earlier in their article (see Fig. 2.18b, their Fig. 2.2). The authors consider the different cases of time-ordering of the emission and absorption events. For the first case,

> [the connections] (5.1), (5.2), (5.3) and (5.4) are represented by the diagrams of [Fig. 2.18a]. One will at once notice that this is the very figure of self-energy (Fig. 2 [should read Fig. 5!, my Fig. 2.18b]). Indeed this process is to be interpreted as the self-energy or the mass increase of the intermediate electron. (RadReacIIIb, p. 131)

Fig. 2.17 The possible connections that could contribute to divergent corrections (RadReacIIIb, p. 130)

(I)

(......k, j^*, j, l.........	......j^*, j ....)	(5·1)
(......k, j^*, j, l.........	......Coulomb......)	(5·2)
(......k, j, j^*, l.........	......j^*, j ....)	(5·3)
(......k, j, j^*, l.........	......Coulomb......)	(5·4)

(II)

(......k, j^*, g, j, l ...	......j^*, j ....)	(5·5)
(......k, j^*, g, j, l ...	......Coulomb......)	(5·6)
(......k, j, g, j^*, l ...	......j^*, j ....)	(5·7)
(......k, j, g, j^*, l ...	......Coulomb......)	(5·8)

(III)

(......k, j^*, j, l.........	......j^*, g, j ......)	(5·9)
(......k, j, j^*, l.........	......j^*, g, j ......)	(5·10)

(IV)

(......k, j^*, g, j, l ...	......j^*, g, j ......)	(5·11)
(......k, j, g, j^*, l ...	......j^*, g, j ......)	(5·12)

For a given process the quantitative contribution is a term in the perturbative expansion and has a divergent part. The divergent parts of all four terms can be eliminated by introducing a *single* correction term $-\delta m\, u^* \beta u$ into the interaction Hamiltonian.

The remaining three groups of processes that lead to divergences contain pairs of terms that all cancel each other out because the processes of each pair differ from each other only by an inversion of the order of emission and absorption of the virtual photon, which in the quantitative term manifests itself in a change of sign.

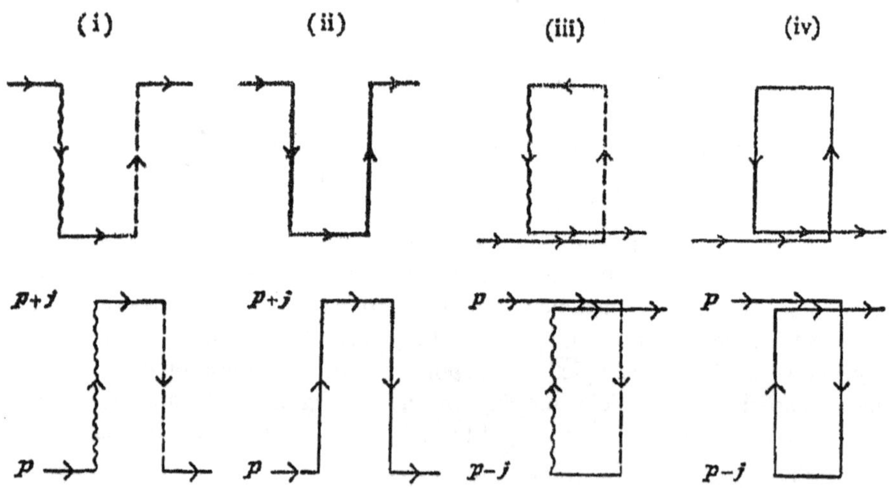

Fig. 2.18 A comparison of transition diagrams for divergent radiative corrections and self-energy (a) Transition diagrams (Figure 8 in RadReacIIIb, p. 131) corresponding to the four divergent sequences of Fig. 2.17. (b) Transition diagrams representing processes that contribute to the self-energy of an electron (Figure 5 in RadReacIIIa, p. 65)

2.4.2 Polarization-Type Divergences

The essential instances of the polarization-type divergences arise from the island-like corrections that share one photon with the original process (see Fig. 2.16c); this type of correction is labelled $[P, I]$. The corrections can be obtained from a new transition diagram constructed from the original transition diagram of the uncorrected process:

> Take an arbitrary connection of the original (uncorrected) process and fix your eyes upon its certain step, the emission of a photon k, say. By virtue of the diagram method we can easily construct $[P, I]$ corrections with regard to this element in the following manner. (RadReacIIIb, p. 133)

A first possibility of an island-like correction is obtained by adding a closed rectangle (the "island") before the square well of the uncorrected process with respect to field time (the horizontal axis) and replacing the label k^*, which indicates the emission of a photon with momentum k in the transition diagram of the uncorrected process, with $(-k)$, the absorption of a photon with like momentum. There are two possibilities for assigning the direction of the arrows to the lines forming the closed rectangle: either clockwise or anticlockwise. The authors' instructions read as follows and the transition diagrams of the uncorrected process and the two possible corrections are shown in Fig. 2.19.

> Put an island-like connection consisting of two elements k^* and $(-k^*)$ into any position before the original $t(k^*)$ [i. e. field time of emission] and at the same time replace the element k^* on the main electron-line by $(-k)$, which of course does not alter the configuration of the original diagram for the uncorrected connection. There are two ways of building the closed ilne [line] according to the direction of the proper time, as will be seen from [Figs. 2.19b and 2.19c]. (RadReacIIIb, p. 133)

After these first two types of island-like corrections, the authors introduce four additional types that differ from the first two by the position of the island in field time. Thus the "configurations of the $[P, I]$ corrections [are] clarified" (p. 135). The position of the island, however, is not yet completely determined. In the first case,

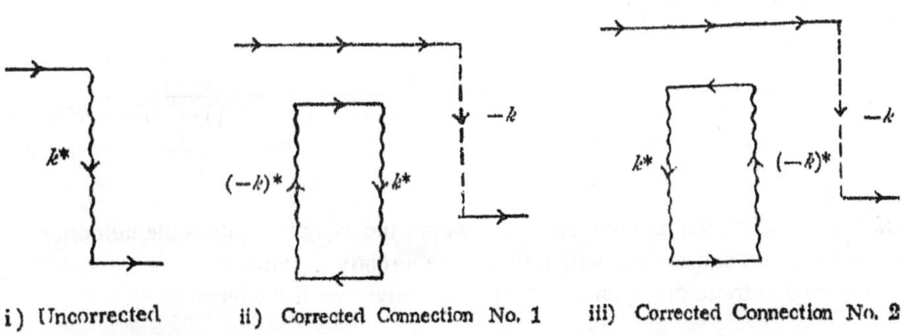

Fig. 2.19 An uncorrected process and two possible polarization corrections (Figures 10(i), (ii), (iii) in RadReacIIIb, pp. 133/134)

for instance, it is only specified that the island lies before the square well to which it is a correction with respect to field time. As long as one meets this condition, one can choose the position and width of the island. Depending on the choice that is made, the degree of divergence of the corresponding quantitative expression differs. If an island lies entirely in one intermediate state, in other words, if no square well of the original process lies between its two vertical elements, the divergences are quadratic, which is the strongest type of divergence that can occur in this context. For islands that extend over two intermediate states of the original process, the divergences are linear; for islands extending over three intermediate states, the divergences are logarithmic—the weakest type of divergence. If the island extends over more than three intermediate states, that is, between its two vertical elements lie two or more square wells of the original process, the correction converges.

Quadratic divergences are denoted by $[S]$ if the island lies between the $(S-1)$th and the Sth square well of the original process. Logarithmic divergences can stem either from islands in one intermediate state, denoted by $\{S\}$, or from islands over two intermediate states, $\{S, S+1\}$. Linear divergences can arise from these two types of islands—these divergences are denoted by (S), $(S, S+1)$—or from islands extending over three intermediate states, $(S, S+1, S+2)$. All quantitative expressions for the corrections have the same numerator, which is reflected in the fact that the transition diagrams all contain an island separated from the rest of the diagram:

> As is suggested by the closed configuration, the polarization correction makes up a separate "Spur" in the numerator of the transition matrix element and this is common to all the connections above described. The only difference between them lies in the energy-denominator to which the order in the field time is essential. (RadReacIIIb, p. 136)

By carrying out an "elementary calculation" (p. 136), the authors obtain the quantitative expressions for the corrections, which they have classified on the basis of their transition diagrams, for instance:

$$[n] = -\left(\frac{4e^2}{3\pi k}\int r\,dr\right)
\frac{H_Q}{(E_0 - E_1)\cdots(E_0 - E_{n-1})(E_0 - E_n)^2(E_0 - E_{n+1})\cdots(E_0 - E_{Q-1})}, \tag{2.7}$$

where r and k are the momenta of the photons and H_Q abbreviates the numerator of the uncorrected expression, which they have already calculated.

All the quadratic divergences cancel each other out if the term

$$\frac{e^2}{3\pi^2}\int r\,dr \int A^2(X)\,dV$$

is added to the interaction Hamiltonian, "because the latter implies processes just equivalent to the above mentioned ones".[18] $A(X)$ is the transverse part of the electromagnetic potential at the point X.[19]

As regards the linear divergences, the authors can identify groups of terms that cancel each other out. Using their graphical classification scheme, they see that none of the linearly divergent terms remains:

Between the terms we find the following relations:

$$\begin{aligned}
\{0\} + \frac{1}{2}\{0, 1\} &= 0 \\
\frac{1}{2}\{0, 1\} + \{1\}1\frac{1}{2}\{1, 2\} &= 0 \\
&\dots \\
\frac{1}{2}\{n-1, n\}\{n\} + \frac{1}{2}\{n, n+1\} &= 0 \\
&\dots \\
\frac{1}{2}\{Q-1, Q\} + \{Q\} &= 0
\end{aligned} \tag{2.8}$$

Therefore, when summed up the linear divergences entirely disappear.

Similarly, groups of logarithmic divergences cancel each other out; however, one divergent term remains:

$$\begin{aligned}
(0) &= 0 \\
(1) + (0, 1) &= 0 \\
(2) + (1, 2) + (0, 1, 2) &= 0 \\
&\dots \\
(n-1) + (n-2, n-1) + (n-3, n-2, n-1) &= 0 \\
(n) + (n-1, n) + (n-2, n-1, n) & \\
&= -\left(\frac{e^2}{3\pi}\int\frac{dr}{r}\right)\frac{H_Q}{(E_0 - E_1)\dots(E_0 - E_{Q-1})} \\
&\dots \\
(n+1) + (n, n+1) + (n-1, n, n+1) &= 0 \\
&\dots \\
(Q) + (Q-1, Q) + (Q-2, Q-1, Q) &= 0.
\end{aligned} \tag{2.9}$$

[18] RadReacIIIb, p. 137.

[19] RadReacIIIa, p. 61.

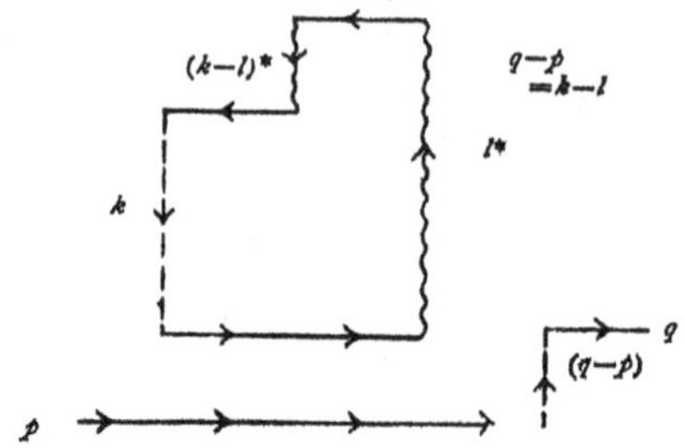

Fig. 2.20 The $[P, II]$ correction to Compton scattering (Figure 14 in RadReacIIIb, p. 140; cf. Figure 17). The line going from p to q with the square step appears discontinuous only because of the loss in quality in picture reproduction

The remaining divergence comes from the integral $(e^2/3\pi)\int(dr/r)$. It is cancelled out by a change in the elementary charge parameter, which describes the coupling between electrons and the electromagnetic field. The observed coupling does not correspond to the natural constant e but to a charge that deviates from that value by a factor proportional to the mentioned diverging integral: $\delta e \sim -(e^2/3\pi)\int(dr/r)$. The effect of "charge renormalization" (p. 140) therefore cancels out the remaining linear divergence.

The $[P, I]$ corrections considered so far are the island-corrections to a square well in the transition diagram of the original process. The well is still present in the corrected diagram but it has been turned from an emission into an absorption. With $[P, II]$ corrections the original diagram appears modified in the corrected diagram. To obtain a $[P, II]$ correction, one island is constructed out of the two square wells of the original process. A typical example is a correction to Compton scattering. The transition diagram of the original process is shown in Fig. 2.14. The corrected transition diagram is shown in Fig. 2.20 and the corresponding Feynman diagram in Fig. 2.21.

The quantitative contribution of the correction is zero due to Furry's theorem, which states that any odd-order contribution in the perturbative expansion vanishes. The Feynman diagrams of these contributions contain a closed polygon of electron lines with an odd number of vertices (three in the case shown in Fig. 2.21), cf. Section 6.6.1.

The diagram in Fig. 2.20 is immediately interpreted as the representation of a physical process:

> From this diagram one will at once recognize that the island-connection represents a process of splitting up of a photon $\tilde{k}$ into two, $\tilde{l}$ and $\widetilde{k-l}$. (RadReacIIIb, p. 140)

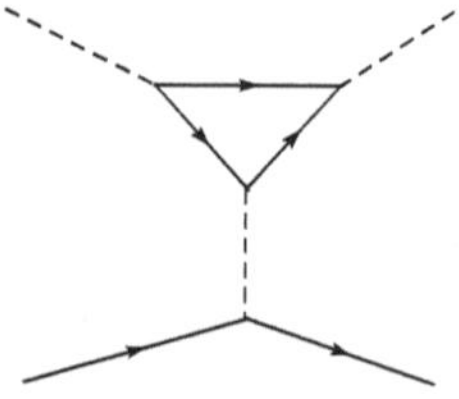

Fig. 2.21 The Feynman diagram (by A. W.) corresponding to the $[P, II]$ correction to Compton scattering shown in Fig. 2.20

Physicists accustomed to the representation of these processes by Feynman diagrams (see Fig. 2.21) would probably have some difficulty interpreting the transition diagram, not "at once" seeing that it shows a photon split into two, while in the corresponding Feynman diagram this would be evident for them.

Furthermore, the $[P, III]$ corrections contain an element that would, according to the "rules" and "notation", represent a "splitting" of a photon: a process that actually does not occur, although:

> We may remark by the way that $[P, III]$ corrections, when separately considered, should not only converge but also vanish; otherwise a photon could spontaneously split up into three photons. (RadReacIIIb, p. 141)

The authors do not explain why they would find the splitting of a photon into three photons impossible. However, it is certainly a shortcoming of their diagrammatic method that no clear criterion is given to sort out the diagrams to which no physical process corresponds.

Thanks to their classification of the correction terms, the authors know that now they have completed their task:

> It is from the outset obvious that all further corrections $[P, IV]$, $[P, V] \ldots [P, Q]$ cannot contribute to divergence since the energy-denominator will contain more than 4 $|r|$s. ($E_r \sim |r|$ is the energy of a virtual electron, and one has to integrate with respect to $|r|$). Moreover all $[P, S]$ corrections with even S do vanish, if the presence of other electrons is not taken into account, because of symmetry property. (RadReacIIIb, p. 141)

Thus they have showed that changing the interaction Hamiltonian proposed at the beginning of the third part of their article (RadReacIIIa, p. 60) is indeed sufficient to remove all the divergent quantities from the theory: the mass-type divergences discussed in the previous section are eliminated by the introduction of a counter-term containing a mass parameter, $\delta m - \delta m\, u^* \beta u$. The polarization-type divergences are either cancelled out by other members of the same subgroup or removed by counter-terms which contain the charge parameter e, either in expressions such as "$\delta e \sim -(e^2/3\pi) \int (dr/r)$" or "$\frac{e^2}{3\pi^2} \int r\, dr \int A^2(X)\, dV$" (equations 4.9 and 4.13, RadReacII, pp. 415/416).

2.4.3 Bookkeeping Versus Isolating the Infinities

It is striking how, despite knowing so little of the developments taking place in Europe and the United States, these Japanese physicists nevertheless arrive at remarkably similar conclusions concerning the removal of QED's uninterpretable infinities. However, their representations of the processes and the associated classification of the problematic terms differ significantly from Dyson's method employing Feynman diagrams. In Chapter 6 we will see in more detail what Dyson's method consists of and the main function of Feynman diagrams. The method that would eventually become established is based, to a large extent, on Dyson's work and has less in common with the Japanese method.[20] The question therefore arises as to the reasons for this preference.

In Section 2.3 I pointed out a possibly crucial difference in the diagrams of the two methods: only Feynman diagrams show whether an emission and an absorption process involve the same photon. Moreover, in contrast to the method of the Japanese physicists, Dyson's procedure yields finite results to *any* order of the perturbative expansion, not just to the first order.[21] Indeed, Dyson's identification of the "self-energy" and "vertex" parts of Feynman diagrams enables him to define an iterative procedure to eliminate the infinities to higher and higher orders.

How much the advantages of Feynman diagrams over the alternative Japanese representations weighed in the actual evaluation of the two methods remains to be investigated. It seems, however, that these advantages made Feynman diagrams more appropriate for the task of justifiably and satisfactorily removing the infinities. More effort needed to be expended on bookkeeping all the terms in the Japanese method in order to assure that they were all taken into account. Indeed, to this end the authors even introduced a "representation by sequences"[22] in addition to the transition diagrams. Dyson, however, is able to "isolate the virus"[23] more effectively than the Japanese physicists.

---

[20] Cf., for instance, Mills (1993).

[21] SM, abstract; RadReacIIIa, p. 61.

[22] RadReacIIIa, p. 65.

[23] Mills 1993, p. 77.

Chapter 3
Quantum Mechanics Without a Hamiltonian Operator

But what does the quantum mechanician do if he cannot be provided with a Hamilton operator? He goes on strike.
Heinrich Leutwyler
Lecture notes 'Elektrodynamik und Optik' (p. 21)
Translation A. W.

3.1 The Main Difficulties of QED

In the early 1940s, when Feynman was a graduate student, one of the most pressing problems facing theoretical physicists of the time was the fact that infinite and, therefore, uninterpretable quantities arose from some of the principles of electrodynamics—in both classical electrodynamics as well as in the early attempts to establish a quantum theory.[1] Roughly, the problem was as follows: an *extended* electron could not be stable because the charges distributed over its volume would repel each other. On the other hand, a *point-like* electron would imply that the energy contained in the electron's electrostatic field would be infinite, since the field energy e/r diverges in the limit of vanishing radius r of the spatial distribution of the charge e.[2] In classical electrodynamics, the difficulties of divergences had been known for some time, and it had been hoped that quantizing the theory would eliminate them. The other strategy was that one should first remove the infinite quantities in the classical theory before one should even attempt to quantize the theory.[3]

It is in this context that Feynman wrote his PhD thesis, with the removal of the divergences in electrodynamics being his superordinate objective.[4] In his thesis, he adopts the second strategy mentioned above of first trying to establish a divergence-free classical theory of electrodynamics and then proceeding to quantize it. Indeed, together with his supervisor John A. Wheeler, Feynman had already developed an

---

[1] For a comprehensive account of the history of QED, see Schweber (1994).

[2] See, e. g., Shapere (1984); Frisch (2005).

[3] See, e. g., Dirac (1938); Frenkel (1925).

[4] Feynman (2005), Introduction.

A. Wüthrich, *The Genesis of Feynman Diagrams*, Archimedes 26,
DOI 10.1007/978-90-481-9228-1_3, © Springer Science+Business Media B.V. 2010

alternative theory of electrodynamics with the desired feature that awaited quantization by the time he had started working on his thesis.[5]

The standard procedure for quantizing a classical theory was to interpret the classical Hamiltonian function as an operator in a Hilbert space of state vectors. This operator would then determine the time evolution of the quantized system described by a certain state vector. The problem with quantizing the Wheeler–Feynman theory of electrodynamics was that it could not be formulated by specifying a Hamiltonian function. Therefore, a method was needed to quantize physical systems, the classical description of which could not be given by a Hamiltonian function.[6]

3.2 How the Classical Action Constructs the Quantum Theory

The solution to the problem of quantizing a system without a Hamiltonian consists in recognizing that the Hamiltonian is not the only classical quantity that can be used to determine the dynamics of a corresponding quantum system. Not only can the classical *action* do so; it also has the advantage that it can construct the quantum theory if the classical theory does not provide a Hamiltonian.

3.2.1 Using the Lagrangian to Represent the Wave Function

Feynman bases his quantization procedure on Paul Dirac's considerations on how to construct a quantum theory out of the Lagrangian formulation of classical mechanics, rather than out of the Hamiltonian formulation. Systems that can be described by a Lagrangian can also be described by a Hamiltonian: the two descriptions are related by so-called Legendre transformations. This also means that a system that has no Hamiltonian has no Lagrangian either: if it did have one, one could construct a Hamiltonian of the system using Legendre transformations, which contradicts the hypothesis that the system has no Hamiltonian. Therefore, Dirac's quantization of a Lagrangian formulation does not solve Feynman's problem of how to quantize a system without a Hamiltonian. However, it does enable him, as we will see in a moment, to generalize to systems without a Hamiltonian.

In his thesis, Feynman quotes from Dirac's 1935 *The Principles of Quantum Mechanics*,[7] over almost two pages, starting with:

> From the analogy between classical and quantum contact transformation ... we see that $(q'_t|q'_T)$ corresponds in the classical theory to $e^{iS/\hbar}$ where S is Hamilton's principal function [i. e. the action] for the time interval T to t, equal to the time-integral of the Lagrangian L,

---

[5] Only a summary of a presentation of Wheeler and Feynman's theory had been published by the time Feynman started working on his thesis (see "Minutes of the Cambridge, Massachusetts, Meeting, February 21 and 22, 1941", p. 683). The published accounts of the theory are Wheeler and Feynman (1945, 1949).

[6] Feynman (2005), Introduction.

[7] Dirac (1935, pp. 124–126).

$$S = \int_T^t L\, dt. \tag{3.1}$$

Taking an infinitesimal time interval t to $t + \delta t$, we see that $\left(q'_{t+\delta t}|q'_t\right)$ corresponds to $e^{\frac{iL\delta t}{\hbar}}$. This result gives probably the most fundamental quantum analogue for the classical Lagrangian function.[8]

In these passages, Dirac is rather vague about the exact nature of the relationship between the exponential $e^{\frac{iL\delta t}{\hbar}}$ of the Lagrange function and the function $\left(q'_{t+\delta t}|q'_t\right)$ that relates the state descriptions in terms of the coordinates q at two different times. In his thesis, Feynman shows that $(q_{t+\delta t}|q_t)$ not only "corresponds to" $e^{\frac{iL\delta t}{\hbar}}$ but that it is actually *proportional* to it. According to Feynman's reminiscences, he discovered this after a beer party in Princeton, when Herbert Jehle, a visiting scholar, called Feynman's attention to Dirac's work on the role of the Lagrangian in quantum mechanics. Feynman, so the story goes, was convinced that Dirac just meant "is proportional to" when he wrote "corresponds to", but Jehle insisted that Feynman had stumbled upon something that nobody else knew: "I [Feynman] didn't realize—I was only trying to interpret Dirac—but he [Jehle] realized that I had discovered something that wasn't known. He said, 'You Americans, always trying to find a use [for] something! That is a way [to] discover new things.'"[9]

However, in a section preceding the one from which Feynman quotes, and in an article predating the publication of the book, Dirac himself states that the two quantities are equal.[10] Feynman most probably simply overlooked these passages or felt the need to work out the relationship in more detail, using his own framework.

Feynman justifies the claim that these two functions are proportional by using their proportionality to derive Schrödinger's equation from a general relationship between the wave function ψ at two infinitesimally close points in time, t and $t + \delta t$[11]:

$$\psi\left(q'_{t+\delta t}, t + \delta t\right) = \int \left(q'_{t+\delta t}|q'_t\right) \psi\left(q'_t, t\right) \sqrt{g\left(q'_t\right)}\, dq'_t, \tag{3.2}$$

where $\sqrt{g\left(q'_t\right)}\, dq'_t$ is the volume element for integration in the space of the variable q'_t.

The wave function is one way of specifying the state vector, and Schrödinger's equation describes the (non-relativistic) time evolution of the wave function. If $\left(q'_{t+\delta t}|q'_t\right)$ is indeed proportional to $e^{\frac{i\delta t}{\hbar} L\left(\frac{q'_{t+\delta t} - q'_t}{\delta t}, q'_{t+\delta t}\right)}$, the above equation is equivalent to

[8] Quoted in Feynman (2005, p. 26).

[9] Weiner (1966b, p. 148–149), reprinted with permission. Copyright 1966, American Institute of Physics.

[10] Dirac (1935), equation 68, p. 113; Dirac (1933), equation 7.

[11] Feynman (2005, p. 28).

$$\psi\left(q'_{t+\delta t}, t+\delta t\right) = \int e^{\frac{i\delta t}{\hbar} L\left(\frac{q'_{t+\delta t} - q'_t}{\delta t}, q'_{t+\delta t}\right)} \psi\left(q'_t, t\right) \frac{\sqrt{g}\, dq'_t}{A(\delta t)}, \tag{3.3}$$

where $A(\delta t)$ is a normalization constant.[12]

This relationship (3.3) between the values of the wave function at different times entails Schrödinger's equation. Feynman proves this statement in the special case of a particle of mass m moving in one dimension under the influence of a force described by the potential $V(x)$. In this case, the Lagrangian is $L = \frac{1}{2}m\left(\frac{x-y}{\varepsilon}\right)^2 - V(x)$ and we obtain (Feynman now uses ε instead of δt)

$$\psi(x, t+\varepsilon) = \int e^{\frac{i\varepsilon}{\hbar}\left\{\frac{m}{2}\left(\frac{x-y}{\varepsilon}\right)^2 - V(x)\right\}} \psi(y, t)\frac{dy}{A}. \tag{3.4}$$

By using a substitution of variables and a Taylor expansion, Feynman shows that the integral equation for the time evolution of the wave function is equivalent to the Schrödinger equation:

$$-\frac{\hbar}{i}\frac{\partial\psi}{\partial t} = -\frac{\hbar^2}{2m}\frac{\partial^2\psi}{\partial x^2} + V(x)\psi. \tag{3.5}$$

Therefore, the two equations "serv[e] the same function" (Feynman 2005, p. 29). Like the Schrödinger equation, the integral in (3.3) containing the classical Lagrangian determines the time evolution of a quantum mechanical state.

To have a relationship between the wave function at two times infinitesimally close to each other does not, however, solve Feynman's problems, which include calculating the probability of transitions from one state at time t_0 to another state at a later time, denoted by T, with no restriction on the difference $T - t_0$. To extend his result (3.3), Feynman divides the interval $[t_0, T]$ into "a very large number of small time intervals t_0 to t_1, t_1 to t_2, ..., t_m to T"[13] and applies the formula (3.3) to each time interval. In the passage quoted by Feynman, Dirac applies a similar division of finite time intervals to expressions such as $\left(q'_T | q'_{t_0}\right)$.[14] Then Feynman formulates a proof "by induction"[15] for the case that the relationship is valid for arbitrarily long intervals of time.

Feynman's inductive construction of the wave function goes as follows: use the abbreviation q_i for q_{t_i}, then the relationship (3.3) is approximately valid for the small but still finite interval from t_i to t_{i+1}, that is

[12] Feynman (2005, p. 28).

[13] Feynman (2005, p. 31).

[14] Dirac (1935), equation 24, p. 125.

[15] Feynman (2005, p. 31).

$$\psi(q_{i+1}, t_{i+1}) \approx \int e^{\frac{i}{\hbar} L\left(\frac{q_{i+1}-q_i}{t_{i+1}-t_i}, q_{i+1}\right)(t_{i+1}-t_i)} \psi(q_i, t_i) \frac{\sqrt{g(q_i)}\, dq_i}{A(t_{i+1}-t_i)}. \tag{3.6}$$

By applying the same relationship to $\psi(q_i, t_i)$ and then again to $\psi(q_{i-1}, t_{i-1})$ in the newly obtained equation, and so forth, Feynman obtains (where $Q = q_{m+1}$, $T = t_{m+1}$)[16]

$$\psi(Q, T) \approx \iint \dots \int \exp\left\{\frac{i}{\hbar} \sum_{i=0}^{m} \left[L\left(\frac{q_{i+1}-q_i}{t_{i+1}-t_i}, q_{i+1}\right)(t_{i+1}-t_i)\right]\right\} \times \psi(q_0, t_0) \frac{\sqrt{g_0}\, dq_0 \sqrt{g_1}\, dq_1 \dots \sqrt{g_{[m]}}\, dq_m}{A(t_1-t_0)A(t_2-t_1)\dots A(T-t_m)}. \tag{3.7}$$

In the limit of infinitesimally small intervals, the approximate relationship is supposed to turn into an equality and the sum in the exponential could supposedly be replaced with an appropriate integral. The problem, however, is finding a definition for the integration prescription $dq_0\, dq_1 \dots dq_m$ in the limit $m \to \infty$, that is, where integrations over infinitely many variables, infinitely close to each other, should be performed. Feynman is unable to define such a procedure precisely and to justify rigorously replacing the sum with an integration. He does not consider himself to be in a position to answer the "difficult mathematical question as to the conditions under which the limiting process of subdividing the time scale, required by equations such as (3.7) actually converges".[17] He only gives a qualitative description of the limiting process and points out the similarity between the sum in the exponential and a definition of the corresponding integral:

> In the limit as we take finer and finer subdivisions of the interval t_0 to T and thus make an ever increasing number of successive integrations, the expression on the right side of [3.7] becomes equal to $\psi(Q, T)$. The sum in the exponential resembles $\int_{t_0}^{T} L(\dot{q}, q)\, dt$ with the integral written as a Riemann sum. (Feynman 2005a, p. 31)

3.2.2 Application to Systems Without a Lagrangian Function

Supposing that the aforementioned sum can indeed be replaced with the corresponding integral, then the exponential contains the classical *action*, $S = \int_{t_0}^{T} L(\dot{q}, q)\, dt$, along the path approximated by the sequence (q_0, t_0), (q_1, t_1), ..., (Q, T). This feature of expressions such as (3.7) will eventually allow Feynman to transgress the limitations of the standard quantization procedure.

All the predictions made by a quantum theory can be deduced from the matrix elements of its operators associated with measuring observable quantities. The matrix elements are given by the scalar product of two state vectors, one of which is

[16] Feynman (2005), equation 47.

[17] Feynman (2005, p. 69).

acted upon by the operator under consideration. According to the standard description of wave functions, the matrix element of an operator F acting at time t_0 on a system that at time T_2 was in a state described by the wave function χ and at time T_1 was in a state described by the wave function ψ is[18]

$$\langle \chi | F | \psi \rangle = \int \bar{\chi} e^{\frac{i}{\hbar} H (T_2 - t_0)} F e^{\frac{i}{\hbar}(t_0 - T_1) H} \psi \, d\mathrm{Vol}, \tag{3.8}$$

where $\bar{\chi}$ is obtained by a complex conjugation from χ, dVol is the volume element for integration using the appropriate variables (which Feynman does not explicitly give) and H is the Hamiltonian of the system. H appears here because, according to the standard description, it determines the time evolution of the wave function. For instance, the wave function ψ_{t_2} at time t_2 of a system that at time t_1 was in a state described by ψ_{t_1} is[19]

$$\psi_{t_2} = e^{\frac{i}{\hbar} H (t_2 - t_1)} \psi_{t_1}. \tag{3.9}$$

Using Dirac's work as his basis, Feynman has already found how the time evolution is given in terms of the Lagrangian L of the system instead of its Hamiltonian (see (3.7)). Therefore, the matrix elements are given as follows[20]:

$$\begin{aligned} \langle \chi | F | \psi \rangle \approx & \iint \ldots \int \chi^*(q_{m+1}, t_{m+1}) \\ & \times \exp \left\{ \frac{i}{\hbar} \sum_{i=-m'}^{m} \left[L \left(\frac{q_{i+1} - q_i}{t_{i+1} - t_i}, q_{i+1} \right) (t_{i+1} - t_i) \right] \right\} F \\ & \times \psi(q_{-m'}, t_{-m'}) \frac{\sqrt{g}\, dq_{m+1} \ldots \sqrt{g}\, dq_0 \sqrt{g} dq_{-1} \ldots \sqrt{g}\, dq_{-m'}}{A(t_{m+1} - t_m) \ldots A(t_0 - t_{-1}) \ldots A(t_{-m'+1} - t_{-m'})}. \end{aligned} \tag{3.10}$$

However, Feynman's goal is to find a quantum theory for classical systems that have no Hamiltonian and, therefore, as mentioned, no Lagrangian. To these systems, equations such as (3.3) and (3.10), containing as they do L, cannot be applied. An example of such a theory is the divergence-free classical electrodynamics of Wheeler and Feynman. How can one determine the time evolution and the matrix elements in the quantized versions of such theories?

Here we see clearly that the main purpose of Feynman's alternative formulation is neither to provide handy mnemonics to be used to calculate cases that can also be described using standard formulations, nor just for the "pleasure in recognizing

[18] Cf. equation 62 in Feynman (2005, p. 39).

[19] Feynman (2005), equation 61.

[20] Cf. equation 50 in Feynman (2005, p. 32).

old things from a new point of view".[21] Rather, Feynman needs the alternative formulation to be able to construct a description of important systems that cannot be described using standard formulations; and standard formulations would require that the Lagrangian be "simple", that is, only a function of coordinates and velocities, which is not always the case:

> What we have been doing so far is no more than to reexpress ordinary quantum mechanics in a somewhat different language. In the next few pages we shall require this altered language in order to describe the generalization we are to make to systems without a simple Lagrangian function of coordinates and velocities. (Feynman 2005, p. 39)

Feynman makes the generalization by divorcing the action from its exclusive definition as the time integral of a Lagrangian, $S = \int_{t_0}^{T} L(\dot{q}, q)\, dt$, thus allowing for more general types of action functions. Classical theories without a Hamiltonian or a Lagrangian might still be formulated by a principle of least action. If one ignores the restriction that the action is the time integral of a Lagrangian, one can then insert the classical action of such theories instead of the Riemann sum of a Lagrangian into (3.3) and (3.10):

> We now make the generalization to the case when the classical action need not be of the form $\mathcal{A} = \int L(\dot{q}, q)\, dt$, but is some other more general functional of $q(\sigma)$. In equation [3.10], as has already been remarked, the phase of the exponential is just $\frac{i}{\hbar} \int L(\dot{q}, q)\, dt$ written as a Riemann sum due to our subdivision of the time into finite, but small, intervals. The obvious suggestion is, then to replace this exponent by $\frac{i}{\hbar}$ times the more general action. The action must of course first be expressed in an approximate way in terms of q_i, t_i, in such a way that as the subdivision becomes finer and finer it more nearly approaches the action expressed as a functional of $q(t)$.[22]

3.3 Interpretation: Microscopic Paths

After the Second World War, a condensed and revised version of Feynman's thesis was published in the *Reviews of Modern Physics*, but Feynman still had not solved all the difficulties concerning defining the integration over infinitely many variables, q, that are infinitely close to each other. In the article, he mentions these difficulties as one of the "inadequacies of the formulation" (title of section 11, RMP48):

> The formulation given here suffers from a serious drawback. The mathematical concepts needed are new. At present, it requires an unnatural and cumbersome subdivision of the time interval to make the meaning of the equations clear. (RMP48, p. 384)

In a footnote (fn. 10), he re-emphasizes the provisional nature of the definition of the new type of integral occurring in his equations: "One feels as Cavalieri must have felt calculating the volume of the pyramid before the invention of the calculus."

However, although Feynman still does not have a precise definition of the mathematical expressions he is using, he can at this point interpret them physically. The

---

[21] RMP48, p. 367.

[22] Feynman (2005, p. 41); Feynman writes $\mathcal{A}$ instead of S for the action.

two "postulates" of RMP48 state that the probability for a certain outcome of a position measurement is determined by "the absolute square of a sum of complex contributions, one from each path in the region", and that "[the] paths contribute equally in magnitude, but the phase of their contribution is the classical action (in units of $\hbar$), i. e., the time integral of the Lagrangian taken along the path" (p. 371).

So, what he described in his thesis as an inductive construction of the wave function is now interpreted as being a sum of contributions from every possible *path* that a particle can take. In equations such as (3.3) and (3.10), the action (or, when available, the time integral of the Lagrangian) is evaluated along a path $q(t)$ of the particle. Since one integrates over the q_is, which mark a sequence approximating the path of the particle, the path is varied in every possible way, which is why Feynman's quantization procedure came to be known as *path integral* quantization.

Neither in his thesis nor in the published article does Feynman give a graphical representation of the idea of a path integral. Only later (in the early 1950s), when he was working on helium does he graphically represent the subdivision process (see number 1, Fig. 3.1). Similar figures are used to introduce Feynman's path integral formulation in today's textbooks (see Fig. 3.2).

Feynman investigates the characteristics of the paths that appear in his alternative formulation and recognizes that they are the result of a familiar physical process, namely *Brownian motion*. This interpretation is a consequence of one of the principal results of his thesis: that the time evolution of the wave function is determined by the classical action. In the article, this result reads as follows:

$$\psi(x_{k+1}, t+\varepsilon) = \int \exp\left[\frac{i}{\hbar} S(x_{k+1}, x_k)\right] \psi(x_k, t)\, dx_k/A. \tag{3.11}$$

As in his thesis, Feynman proves that the above integral equation is equivalent to the Schrödinger equation in the "simple case of a particle moving in one dimension in a potential $V(x)$".[23]

After carefully defining the action $S(x_{k+1}, x_k)$ for one section of a path by performing an adequate approximation and replacing $x_{k+1} - x_k$ with ξ and x_{k+1} with x, Feynman finds that

$$\psi(x, t+\varepsilon) = \int \exp\frac{im\xi^2}{\varepsilon \cdot 2\hbar} \cdot \exp\frac{-i\varepsilon V(x)}{\hbar} \cdot \psi(x-\xi, t)\, d\xi/A. \tag{3.12}$$

Since ε is small, the factor $\exp\frac{im\xi^2}{\varepsilon \cdot 2\hbar}$ oscillates rapidly with ξ, except where ξ is "of order $(\hbar\varepsilon/m)^{\frac{1}{2}}$".[24] Therefore, when one integrates over ξ, only the small values of ξ have any effect on the results of the integration. The value of the wave function $\psi(x, t+\varepsilon)$ is, therefore, essentially determined by the values of the earlier wave

[23] RMP48, p. 374.

[24] RMP48, p. 375.

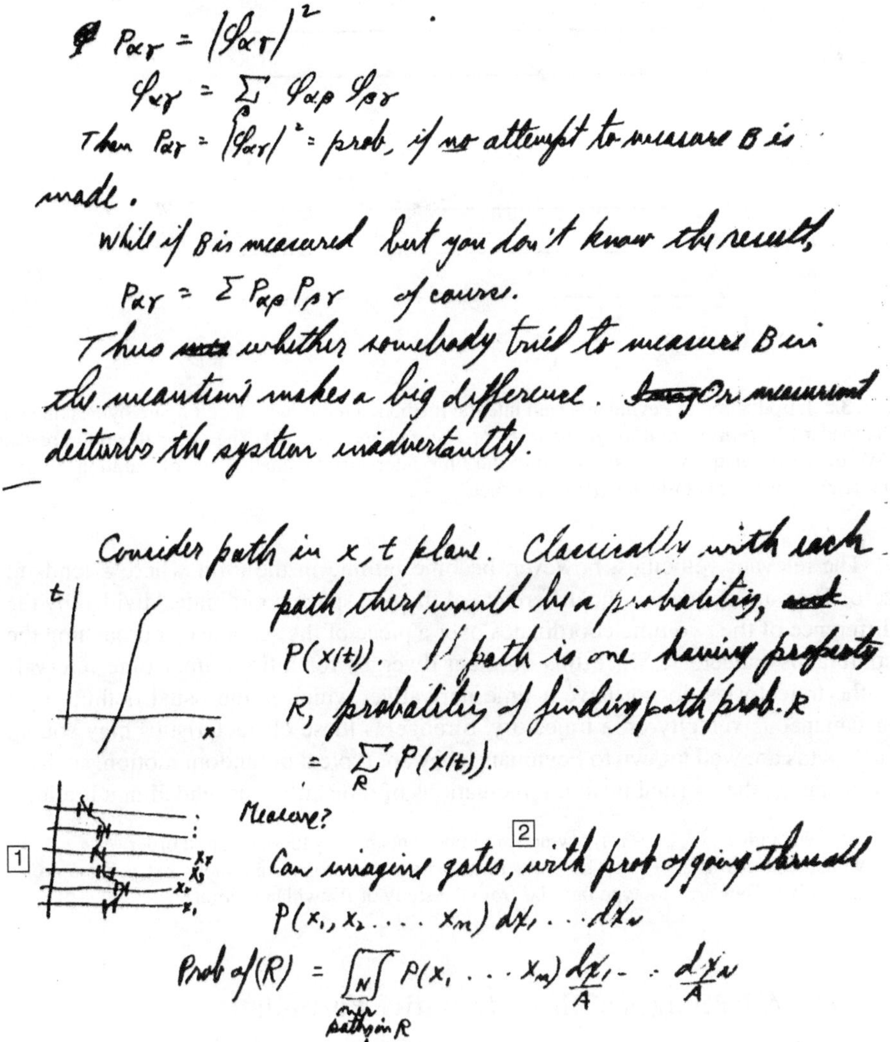

$P_{\alpha\gamma} = |\varphi_{\alpha\gamma}|^2$.

$\varphi_{\alpha\gamma} = \sum_{\beta} \varphi_{\alpha\beta}\, \varphi_{\beta\gamma}$

Then $P_{\alpha\gamma} = |\varphi_{\alpha\gamma}|^2$ = prob. if no attempt to measure B is made.

While if B is measured but you don't know the result,

$P_{\alpha\gamma} = \sum P_{\alpha\beta} P_{\beta\gamma}$ of course.

Thus whether somebody tried to measure B in the meantime makes a big difference. Our measurement disturbs the system inadvertently.

Consider path in x, t plane. Classically with each path there would be a probability, $P(x(t))$. If path is one having property R, probability of finding path prob. R $= \sum_R P(x(t))$.

Measure?

Can imagine gates, with prob of going thru all $P(x_1, x_2, \ldots x_n)\, dx_1 \ldots dx_n$

Prob of (R) $= \int_{[N]}\!\int P(x_1 \ldots x_n) \frac{dx_1}{A} \ldots \frac{dx_n}{A}$ paths in R only.

Fig. 3.1 A graphical representation (*number 1*) of the subdivision of the time scale. Integrating over each x_i corresponds graphically to moving the "gates" (*number 2*) horizontally. Therefore, the integration over each x_i amounts to an integral over all possible paths, early 1950s (*Helium*, folio 12, reprinted with permission of Melanie Jackson Agency, LLC)

function, which are "close to [x] (distant of order $\varepsilon^{\frac{1}{2}}$)". So, when one considers the limit of vanishing ε, the points at which the value of the earlier wave function is relevant remain close to the point x at which the wave function is to be evaluated an instant later. The relevant points thus lie on a continuous trajectory and the integral equation (3.11) for ψ can be replaced with a differential equation.

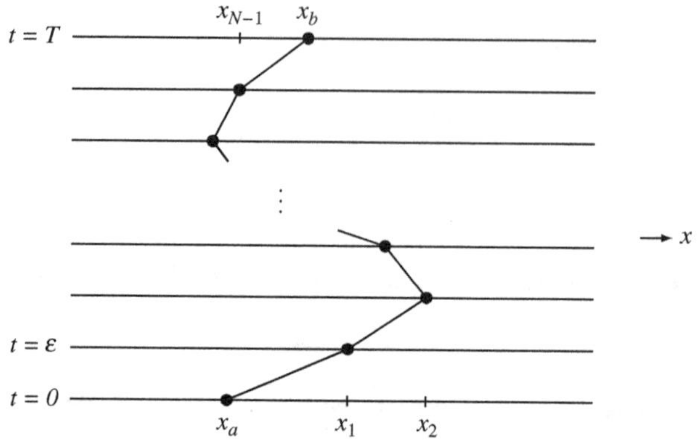

Fig. 3.2 Introduction to Feynman's path integral method: a reproduction of a figure by Peskin and Schroeder (*An Introduction to Quantum Field Theory*, 1995, p. 278). The original caption reads: "We define the path integral by dividing the time interval into small slices of duration ε, then integrating over the coordinate x_k of each slice"

The relevant velocities, however, become infinite in the limit where ε tends to zero. They are defined as the difference of the two space coordinates divided by the difference of the two time coordinates over a piece of the sequence approaching the continuous trajectory. Since this quotient diverges for infinitesimal time intervals ε, the trajectories do not have a time derivative, which is the usual definition of instantaneous velocity on a trajectory. Strange as these characteristics may sound, such paths are well known to Feynman. They are typical of random motion, such as, for instance, the thermal position fluctuations of molecules suspended in a liquid:

> The "velocities", $(x_{k+1} - x_k)/\varepsilon$ which are important are very high, being of order $(\hbar/m\varepsilon)^{\frac{1}{2}}$ which diverges as $\varepsilon \to 0$. The paths involved are, therefore, continuous but possess no derivative. They are of a type familiar from the study of Brownian motion.[25]

3.4 The Advantages of the Alternative Formulation

Feynman felt that if his thesis had contained nothing but an alternative formulation of cases that can also be treated using the conventional formulation, it would have lost much of its attractiveness:

> [He] thought, "No, that's not important at all, because that's exactly equivalent to the regular representation—it's equivalent—so it's just another way to write it, it doesn't add anything. What I really have to get is some contribution, make something different, which is this electrodynamics." (Weiner 1966b, p. 153, reprinted with permission. Copyright 1966, American Institute of Physics)

Feynman wanted to arrive at a divergence-free form of quantum electrodynamics (QED) by way of quantizing the Wheeler–Feynman theory. However, he did not

[25] RMP48, p. 376.

achieve this goal nor could he quantize any other relativistic system using his methods:

> The results of the application of these methods to quantum electrodynamics is [sic] not included in this thesis, but will be reserved for a future time when they shall have been more completely worked out. [...] All of the analysis will apply to non-relativistic systems. The generalization to the relativistic case is not at present known. (Feynman 2005, pp. 5–6)

However, after withdrawing from war-related work some time in the first half of 1942 (see Gleick 1992, pp. 146ff.), he was, after all, able to include the generalization to actions that are not integrals of a Lagrangian in his thesis (see Section 3.2.2). Indeed, it is the possibility of this generalization that constitutes the main advantage of Feynman's alternative formulation. As described above, the recognition that the Lagrangian is only used in the alternative quantization procedure as an intermediate to the action allows one to generalize the procedure to theories that have no Lagrangian but do have an action. The advantage of the alternative formulation is that it suggests how one can make generalizations to hitherto indescribable cases.

But even in the domain of problems that can be represented by conventional methods, Feynman's alternative formulation has a particular advantage. The example he gives is the case of the forced harmonic oscillator, that is, a harmonic oscillator interacting with a (Lagrangian) system. The advantage of using Feynman's alternative formulation for this quite standard problem is that it allows one to describe the dynamics of the (Lagrangian) system without explicitly referring to the motion of the oscillator; only its net effect is taken into account. The net effect can be calculated once and for all, irrespective of the specificities of the interacting system:

> [The problem of the forced harmonic oscillator], when the oscillator is interacting with a Lagrangian system, can of course be handled by the usual methods of quantum mechanics. We shall see, however, that the added power of looking at all times at once, so to speak, which arises in such equations as [3.10], has some advantages. With a wave function, the oscillator and the interacting system are so firmly interlocked, mathematically, that it is hard to study the properties of the oscillator without, at the same time, solving for the motion of the interacting system. We shall be able here, however, to solve that half of the problem which involves the oscillator, without solving the entire problem. (Feynman 2005, p. 55)

From comparable classical cases, Feynman knows that the interaction mediated by an oscillator is not instantaneous but delayed. The description that arises when the oscillator's coordinates are not used thus refers to various different times. As a consequence, it cannot be cast into equations of motion of Hamiltonian form, nor into a Lagrangian action:

> [We] shall expect that the interaction will not be instantaneous, and hence not expressible in Hamiltonian form. [...]
>
> [We] shall expect that the system with the oscillator is not equivalent to the system without the oscillator for all possible motions of the oscillator, but only for those for which some property (e. g., the initial and final position) of the oscillator is fixed. These properties, in the cases discussed, are not properties of the system at just one time, so we will not expect to find the equivalence simply by specifying the state of the oscillator at a certain time, by means of a particular wave function. It is for this reason that the ordinary methods of quantum mechanics do not suffice to solve this problem. (Feynman 2005, p. 62)

However, why would one want to describe the system without using an oscillator? Fermi (1932, p. 131), who Feynman also cites in his thesis (p. 61), had already eliminated the oscillators corresponding to the longitudinal component of the relativistic electromagnetic field, and shown that their net effect is simply the classical (instantaneous) Coulomb interaction energy. For an electron in its own field, this electrostatic energy diverges (see Section 3.1) and is, therefore, one of the sources of the problematic infinities in QED. Removing the oscillators thus produces a description in which one of the problems of the theory—the self-interaction of the electron—is more apparent than in the usual description, which, therefore, gives a better indication as to how to solve the difficulty.

In the RMP48 article, Feynman is slightly more explicit about his reasons for removing the oscillators, stressing that, for doing so in compliance with the theory of relativity, conventional quantum mechanics is an insufficient tool:

> It is true that the oscillators representing the longitudinal waves may be eliminated. The result is instantaneous electrostatic interaction. The electrostatic elimination is very instructive as it shows up the difficulty of self-interaction very distinctly. In fact, it shows it up so clearly that there is no ambiguity in deciding what term is incorrect and should be omitted. This entire process is not relativistically invariant, nor is the omitted term. It would seem to be very desirable if the oscillators, representing transverse waves, could also be eliminated. This presents an almost insurmountable problem in the conventional quantum mechanics. (RMP48, p. 385)

The problem with the traditional method is that one can only refer to one time, whereas the description resulting from the elimination of the oscillators necessarily refers to a number of different times:

> We expect that the motion of a particle a at one time depends upon the motion of b at a previous time, and *vice versa*. A wave function $\psi(x_a, x_b; t)$, however, can only describe the behavior of both particles at one time. There is no way to keep track of what b did in the past in order to determine the behavior of a. The only way is to specify the state of the set of oscillators at t, which serve to "remember" what b (and a) had been doing.
>
> The present formulation permits the solution of the motion of all the oscillators and their complete elimination from the equations describing the particles. (RMP48, p. 385)

Feynman was not the only physicist to come up with a formulation of QED that complies with the theory of relativity; in 1948 Julian Schwinger[26] provided one, too. As in the traditional formulation, he eliminated the longitudinal oscillators and left the transverse oscillators, but in a covariant way. A thorough comparison of Schwinger's and Feynman's formulations is not, however, within the scope of this work.[27]

---

[26] Schwinger (1948b).

[27] For a more extended account of Schwinger's and Feynman's biographies and their contributions to QED, see Schweber (1994, chapters 7 and 8).

3.5 The Unsatisfactory Integration of Spin and Relativity

The results of Feynman's thesis were "non-relativistic throughout". In the last section of the RMP48 article, however, Feynman demonstrates how one can include relativistic systems and particles with spin.

In the conventional formulation of non-relativistic quantum mechanics, phenomena involving electron spin are described by the Pauli equation[28]:

$$-\frac{\hbar}{i}\frac{\partial}{\partial t}\psi(x,t) = H_P\psi(x,t) \tag{3.13}$$

with

$$H_P = \frac{1}{2m} \times \sum_k \sum_l \sigma^{(k)}\sigma^{(l)} \left(\frac{\hbar}{i}\frac{\partial}{\partial x^{(k)}} + \frac{e}{c}A^{(k)}\right)\left(\frac{\hbar}{i}\frac{\partial}{\partial x^{(l)}} + \frac{e}{c}A^{(l)}\right) - eA^{(0)}. \tag{3.14}$$

Feynman could introduce additional terms in his alternative formulation such that it entails the Pauli equation. However, he could not justify the necessary terms by interpreting them physically. The incorporation of spin into his formulation remained "purely formal" (RMP48, p. 387):

> Spin may be included in a formal way. The Pauli spin equation can be obtained in this way: One replaces the vector potential interaction term in $S(x_{i+1}, x_i)$,
>
> $$\frac{e}{2c}(x_{i+1} - x_i) \cdot A(x_i) + \frac{e}{2c}(x_{i+1} - x_i) \cdot A(x_{i+1})$$
>
> [...] by the expression
>
> $$\frac{e}{2c}(\sigma \cdot (x_{i+1} - x_i))(\sigma \cdot A(x_i)) + \frac{e}{2c}(\sigma \cdot A(x_{i+1}))(\sigma \cdot (x_{i+1} - x_i)).$$
>
> (RMP48, pp. 386–387)

In a similar manner Feynman could include relativistic particles without spin (for example, certain mesons), which are described by the Klein–Gordon equation. Feynman could incorporate the content of this equation by defining a four-dimensional path to meet the relativistic requirement that the time variable not be singled out. In the article, Feynman presents a Lagrangian defined in terms of derivatives of these four-dimensional paths. Using this Lagrangian to define the action, Feynman claims to be able to construct a wave function that would automatically satisfy the Klein–Gordon equation.

The Dirac equation combines both spin phenomena and relativity. Feynman mentions that, by modifying the Lagrangian from which the Klein–Gordon equation can

---

[28] Pauli (1933, p. 237).

be derived, the Dirac equation would follow, too. The modification is similar to the aforementioned one that is needed to derive Pauli's equation.

Feynman was not at all pleased with these treatments of spin phenomena and relativistic particles. Tantalizingly, he let the reader know that he was working on a more satisfactory treatment of these two subjects, which was not yet ready for publication:

> These results for spin and relativity are purely formal and add nothing to the understanding of these equations. There are other ways of obtaining the Dirac equation which offer some promise of giving a clearer physical interpretation to that important and beautiful equation. (RMP48, p. 387)

In the next section we will learn what Feynman considers to be a more satisfactory formalization of the Dirac equation as well as discover what he means by "understanding", in contrast to being merely able to "deduce", an equation.

Chapter 4
The Dirac Equation: Feynman's Great Struggle

In this chapter I will show how Feynman used diagrams to represent and modify physical models of electrons (and later their interaction) and how he used these representations to derive quantitative expressions that are "true of the model".[1]

4.1 A Brownian Model of the Electron

From the notes and correspondence of Feynman, which I will now present and describe, we will learn that what Feynman means by "understanding" the Dirac equation is not the study of the mathematical properties of the Dirac equation or the search for an ingenious method of solution required by the application of the equation to complex problems. Rather, Feynman is looking for a physical system, the appropriate description of which would satisfy the Dirac equation.

Since Gregory Breit (1928) and Erwin Schrödinger (1930) it had been held that an electron described by Dirac's equation oscillates at the speed of light around a mean trajectory.[2] At first sight this consequence of Dirac's equation seems at variance with the constraints of the relativity theory to the effect that only massless particles can move at the speed of light; for any massive particle such as the electron, the kinetic energy required for such a motion would be infinite. In *The Principles of Quantum Mechanics*, Paul Dirac[3] also shows how his equation leads to this allegedly absurd consequence and immediately goes on to resolve the apparent contradiction. Dirac demonstrates that the velocity associated with the electron's mean trajectory is smaller than the speed of light and thus consistent with the actual measurements and (non-quantum) relativistic relations of momentum and velocity.

Absurd as the idea of a quivering electron might sound at first, Feynman recognizes that this feature of the electron would fit the extension of his alternative

---

1 Cartwright (1983).

2 Cf. Weisskopf (1939, p. 72); cf. also Feynman and Hibbs (1965), problem 2–6, p. 34, where the problem is presented as being made up for pedagogical purposes only and devoid of any relationship to a real physical situation.

3 Dirac (1935, p. 260).

A. Wüthrich, *The Genesis of Feynman Diagrams*, Archimedes 26,
DOI 10.1007/978-90-481-9228-1_4, © Springer Science+Business Media B.V. 2010

formulation to relativistic systems perfectly. In RMP48, Feynman found that the mainly relevant paths are "of a type familiar from the study of Brownian motion" (see Section 3.3). The quivering motion of the electron, or *Zitterbewegung* as Schrödinger (1930, p. 7) called it, is of that type, too.

In his quest to interpret the Dirac equation, Feynman tries to invert the deductive relationship between the *Zitterbewegung* and the Dirac equation. Based on his alternative formulation of non-relativistic quantum mechanics, he tries to find a description of the quivering electron, such that the Dirac equation necessarily remains valid.

4.1.1 The Dirac Equation in One Dimension

In the following, on the basis of manuscripts in the Feynman Collection held by the Caltech Archives (see the Bibliography, p. 112), I will try to reconstruct Feynman's quest to find an interpretation of the Dirac equation. My reconstruction starts with Feynman reducing the Dirac equation to a special version whereby the wave function depends, with the exception of the time parameter, only on one spatial dimension, instead of the usual three spatial dimensions.

In the one-dimensional version of Dirac's equation, the wave function (or spinor), to which the Dirac equation applies, has only two components instead of the usual four in three spatial dimensions. The one spatial dimension is denoted by z and the two spinor components are ψ_1 and ψ_3. Choosing x for the one spatial dimension and index 2 for the second spin component would have been more natural, and indeed Feynman would later use x to denote the one spatial dimension. Perhaps Feynman's decision to use z for the one spatial dimension and index 3 for the second spin component was influenced by the (quite arbitrary) convention in discussions of spin phenomena in ordinary three-dimensional space of singling out the z axis, that is, the third spatial component, as a reference for measuring spin. Time is denoted, as usual, by the parameter t; μ is the mass of the particle. The one-dimensional Dirac equation for the two components then takes the following form (see Fig. 4.1, number 1):

$$\frac{\partial \psi_1}{\partial t} + \frac{\partial \psi_3}{\partial z} = -i\mu\psi_1, \tag{4.1}$$

$$\frac{\partial \psi_3}{\partial t} + \frac{\partial \psi_1}{\partial z} = +i\mu\psi_3. \tag{4.2}$$

When the interaction of the electron with an electromagnetic field, described by the scalar potential ϕ and vector potential A, is taken into account, the terms containing these potentials have to be added. In one spatial dimension, A has only one component instead of the usual three. As is customary, the electric field is given by the gradient of the scalar potential and the time derivative of the vector potential. In one dimension, this is $E = -\frac{\partial \phi}{\partial x} - \frac{\partial A}{\partial t}$ (see number 2; "E", in the

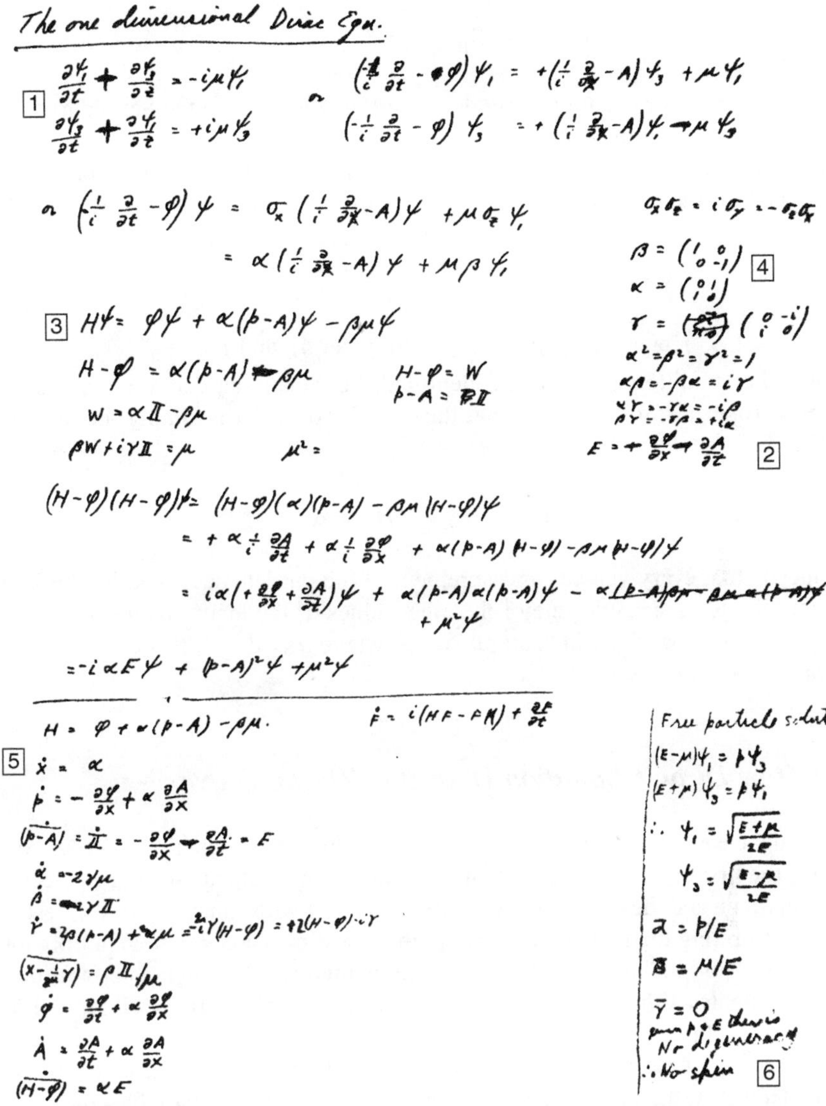

Fig. 4.1 Notes showing the beginning of Feynman's "struggle" to interpret the Dirac equation, ca. 1946 (*Dirac Equation b*, folio 13, reprinted with permission of Melanie Jackson Agency, LLC)

lower half of the right-hand-side column, denotes, instead, the kinetic energy of the particle).

Like the Schrödinger equation, the Dirac equation is a differential equation for the wave function that provides a description of the quantum system under consideration. In the simplest case, the system is a free electron. In the next to simplest case, the electron is interacting with an electromagnetic field. The differential equation and the Hamiltonian, the energy operator of the system, are related by

$$i\frac{\partial\psi}{\partial t} = H\psi, \tag{4.3}$$

as in the Schrödinger equation; only the Hamiltonians differ in the Schrödinger and Dirac equations (Feynman sets $\hbar = 1$).

Feynman brings the one-dimensional Dirac equation into this form in order to be able to read off the Hamiltonian of the system (number 3):

$$H\psi = \phi\psi + \alpha(p - A)\psi - \beta\mu\psi, \tag{4.4}$$

where $\psi = \begin{pmatrix}\psi_1\\ \psi_3\end{pmatrix}$, $\alpha = \begin{pmatrix}0 & 1\\ 1 & 0\end{pmatrix}$, $\beta = \begin{pmatrix}1 & 0\\ 0 & -1\end{pmatrix}$ (number 4), and $p = \frac{1}{i}\frac{\partial}{\partial x}$.

Once the Hamiltonian has been identified, Feynman can derive several of the system's properties, for example, that the velocity operator is simply the matrix α (number 5):

$$\dot{x} = i(Hx - xH) = \alpha. \tag{4.5}$$

This means that every velocity measurement of the particle described by the Dirac equation yields, as a result, one of the eigenvalues of the matrix α, which are ± 1. Since Feynman is working in "natural" units, where $\hbar = c = 1$, this corresponds to plus or minus c, the speed of light.

4.1.2 How Dirac's Equation Describes Zigzag Paths

In the usual case of three spatial dimensions, the Dirac wave function has four components, the first of which describes an electron with spin up, the second an electron with spin down and the third and the fourth a positron with spin up or down, respectively. Interpreting the two components in the one-dimensional case is not obvious as they have "no spin" (see Fig. 4.1, number 6) in this case. Feynman interprets these two components as the description of an electron moving, respectively, to the left or to the right. Interpreted in this way, the wave function helps him to describe the *Zitterbewegung* of the electron, which in one spatial dimension is a simple left-right motion. However, the interpretation can only be given after the variables have been changed and a suitable linear combination of the two components has been formed (see Fig. 4.2).

We can call these "new variables" *light-cone coordinates* (see Louis Kauffman and H. Pierre Noyes (1996)). The new variables are the coordinates in a system rotated by 45° with respect to the x and t axes. Therefore, the axes of the new coordinate system take the same trajectories of a light ray starting or arriving at the origin and moving to the right or left (with respect to the x coordinate). The spinor components are also redefined: they are rotated by 45° with respect to a "spinor" coordinate system that has as its axes the components 1 and 3. The two new variables are u and v (see Fig. 4.2, number 1):

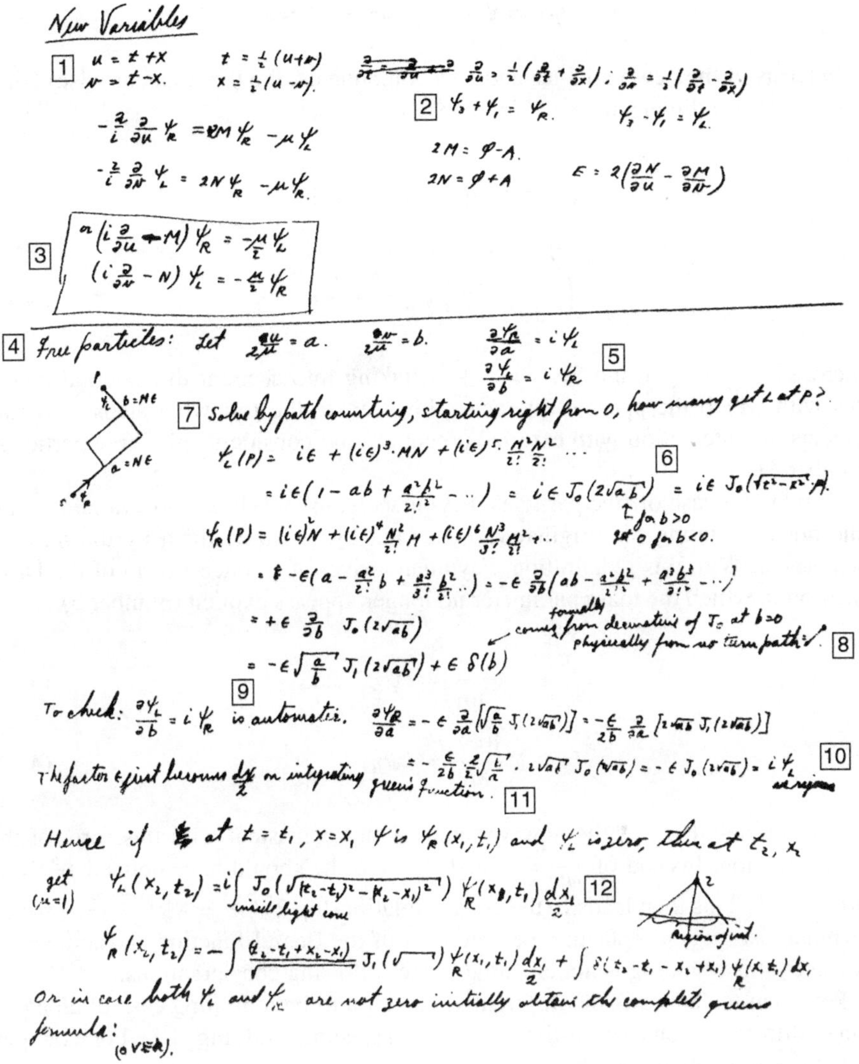

Fig. 4.2 The two components of the wave function in the one-dimensional Dirac equation describe an electron moving, respectively, to the right or left. The equations are expressed in terms of the "new variables" and are solved "by path counting", ca. 1946 (*Dirac Equation b*, folio 10, reprinted with permission of Melanie Jackson Agency, LLC)

$$u = t + x, \tag{4.6}$$

$$v = t - x, \tag{4.7}$$

and the two new spinor components are ψ_R and ψ_L (abbreviations for right and left—why will become clearer as we proceed; see number 2):

$$\psi_3 + \psi_1 = \psi_R, \qquad \psi_3 - \psi_1 = \psi_L. \tag{4.8}$$

In terms of the new variables and new components, the (component-wise) Dirac equation in one dimension reads (number 3):

$$\left(i\frac{\partial}{\partial u} - M\right)\psi_R = \frac{-\mu}{2}\psi_L, \tag{4.9}$$

$$\left(i\frac{\partial}{\partial v} - N\right)\psi_L = \frac{-\mu}{2}\psi_R, \tag{4.10}$$

where $2M = \phi - A$ and $2N = \phi + A$, taking into account the external potentials with which the particle interacts. However, in the following steps Feynman neglects the interaction with external potentials and considers only "free particles" (number 4).

To treat the case of free particles, Feynman again introduces new variables, this time not by rotating the original variables but by rescaling them by the mass of the particle. With this redefinition, Feynman arrives at a concise form of the Dirac equation in which the mass parameter no longer appears explicit (number 5):

$$\frac{\partial\psi_R}{\partial a} = i\psi_L, \tag{4.11}$$

$$\frac{\partial\psi_L}{\partial b} = i\psi_R. \tag{4.12}$$

Feynman's definition of the new variables is not consistent with this form of the Dirac equation. Instead of $\frac{u}{2\mu} = a$ and $\frac{v}{2\mu} = b$, he should have defined $\frac{\mu u}{2} = a$ and $\frac{\mu v}{2} = b$. This then leads to the correct relationship $2\sqrt{ab} = \sqrt{t^2 - x^2}\mu$, which Feynman uses when replacing the argument of the Bessel function J_0 (number 6). However, this rescaling is not essential in the following considerations.

Feynman understands the differential equation to be a limiting case of an equation of differences on a space-time lattice with spacing ε (cf. Fig. 4.3). For example, (4.12) would explicitly read:

$$\lim_{\varepsilon\to 0}\frac{\psi_L(a,b) - \psi_L(a,b-\varepsilon)}{\varepsilon} = i\psi_R(a,b-\varepsilon), \tag{4.13}$$

which can be transformed into

$$\psi_L(a,b) = \psi_L(a,b-\varepsilon) + i\varepsilon\psi_R(a,b-\varepsilon). \tag{4.14}$$

This equation expresses the fact that the component, ψ_L, describing the electron moving to the left on the lattice point (a, b) is determined by the two components of the wave function at the neighbouring point $(a, b-\varepsilon)$: it equals the sum of the "left"

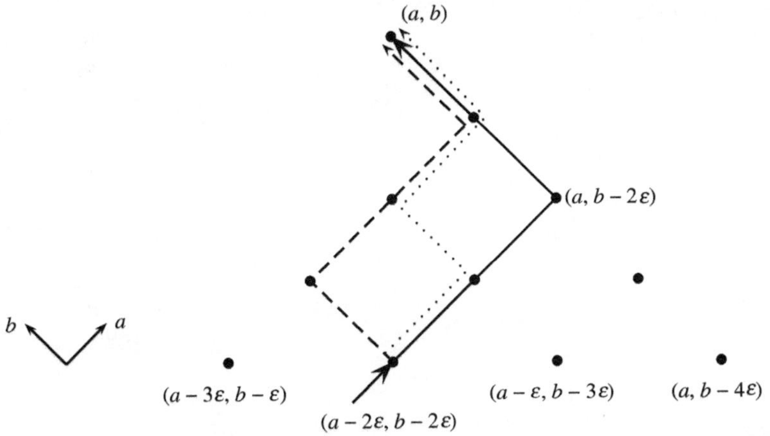

Fig. 4.3 A space-time lattice with coordinate system (a, b) and lattice spacing ε. The solid and dotted lines as well as the dashes show one path with one turn (at $(a, b - 2\varepsilon)$) and two paths with three turns. (The path of dashes turns immediately at the point of origin)

component and the "right" component times $i\varepsilon$. The contribution of the "right" component to the "left" component at the next lattice point represents a change in direction of the electron. Thus, in order to describe the quivering electron by a wave function that would obey the one-dimensional Dirac equation, each change in direction has to be described by a factor $i\varepsilon$.

4.1.3 Path Counting

Dirac's differential equations in Feynman's one-dimensional version provide Feynman with the relationship of the wave function at a certain lattice point to the wave function at a neighbouring point. This result is similar to the observations that Feynman makes in his thesis and in RMP48 that the wave function at a certain space-time point is determined by the wave function one instant earlier and by the classical action. In both texts and in Feynman's notes on the Dirac equation, the relationship can then be applied successively until one arrives at the relationship between the wave function at finitely separated coordinates, instead of at only infinitesimally close coordinates (cf. Section 3.2.1).

As an example, Feynman considers the case where only the "right" component of the initial wave function at the origin of the lattice differs from zero. What then is the "left" component of the wave function at lattice point P?: "starting right from 0, how many get L at P?" Feynman is about to solve this problem by "path counting" (see Fig. 4.2, number 7).

Feynman graphically represents the situation in the diagram (see also left of number 7) reproduced in Fig. 4.4. The point, P, and the origin, 0, are separated

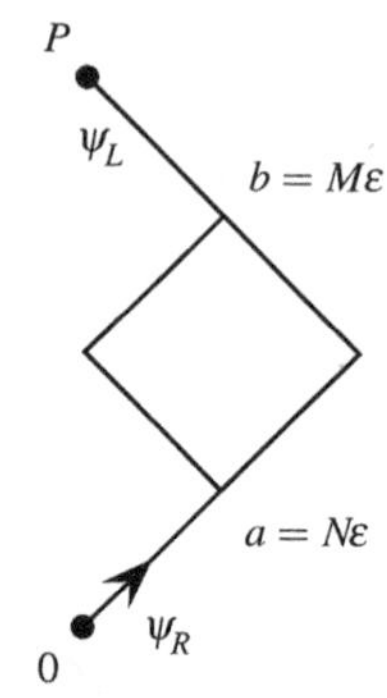

Fig. 4.4 Detail of Fig. 4.2, adapted by the author (A. W.)

by M lattice spacings, ε, in the b coordinate and N lattice spacings in the a coordinate. $M + N$ appropriate iterations of the difference formulae express $\psi_L(a, b)$ as the sum of contributions from lattice points that are $M + N$ lattice spacings away.

Some of the contributions come from the wave function at the origin; this is the contribution which we want to determine. The successive application of the difference formulae defines several paths leading from P to the origin. The difference equation also prescribes that each corner of a particular path is weighted by a factor $i\varepsilon$, as discussed in Section 4.1.2. The problem of determining the contribution from the wave function at the origin then comes down to working out how many of the paths have a particular number of corners.

That the "path counting" procedure is an iterative procedure akin to the "inductive" method in Feynman's thesis becomes more evident when a very coarse lattice is considered. Let us suppose that the lattice is so coarse that the point, P, and the origin are only separated by two lattice spacings in each direction. In that case, P has the coordinates (a, b) and the origin is $(a - 2\varepsilon, b - 2\varepsilon)$ (see Fig. 4.3). Four appropriate iterations of the difference formulae express $\psi_L(a, b)$ as the sum of contributions from lattice points that are connected to P by zigzag paths of a length of four lattice spacings. Two of these contributions come from the particular point $(a - 2\varepsilon, b - 2\varepsilon)$:

$$\begin{aligned}
\psi_L(a,b) &= \psi_L(a, b-\varepsilon) + i\varepsilon\psi_R(a, b-\varepsilon) \\
&= \psi_L(a, b-2\varepsilon) + i\varepsilon\psi_R(a, b-2\varepsilon) \\
&\quad + i\varepsilon[\psi_R(a-\varepsilon, b-\varepsilon) + i\varepsilon\psi_L(a-\varepsilon, b-\varepsilon)] \\
&= \dots \\
&= \dots + i\varepsilon\psi_R(a-2\varepsilon, b-2\varepsilon) + \dots + 2(i\varepsilon)^3\psi_R(a-2\varepsilon, b-2\varepsilon) + \dots
\end{aligned}$$

The two contributions come from one path with one corner, hence a factor $i\varepsilon$, and from two paths with three corners, hence a factor $2(i\varepsilon)^3$.

Returning to the arbitrary fine lattice, which Feynman considers in his notes, there is one path that starts on the "right" of the diagram and ends on the "left" of the diagram (from 0 to P) with one corner, since there is only one point between 0 and P that is connected to 0 by a line that goes up to the right and to P by a line that goes up to the left. This gives a contribution $i\varepsilon$ to $\psi_L(P)$ (see equation below number 7 in Fig. 4.2).

No path (that starts on the "right" of the diagram and ends on the "left" from 0 to P) with an even number of corners exists, since, after an even number of changes in direction, the particle beginning on the right would end again on the right, which gives no contribution to the "left" component of the wave function.

There are $(N-1)(M-1)$ paths with three corners, since there are $(N-1)$ possible lattice points for the first corner and $(M-1)$ for the third, the second corner being determined by the first and third corners.[4] In the continuum limit, that is, in the limit of the vanishing lattice spacing $\lim_{\varepsilon\to 0}$, $M\varepsilon$ and $N\varepsilon$ tend to b and a, respectively. In that limit, $(N-1)(M-1)$ can be replaced with MN, since M and N have to be much larger than 1. The contributions from the paths with three corners are thus $(i\varepsilon)^3 MN$.

By making similar combinatorial considerations, Feynman finds that there are $\frac{M^2}{2!}\frac{N^2}{2!}$ paths with five corners. Accordingly, these give the contribution $(i\varepsilon)^5\frac{M^2}{2!}\frac{N^2}{2!}$. Using the same construction, Feynman obtains $\psi_R(P)$ as the sum of contributions from paths with an even number of corners:

$$\psi_R(P) = (i\varepsilon)^2 N + (i\varepsilon)^4 \frac{N^2}{2!} M + (i\varepsilon)^6 \frac{N^3}{3!}\frac{M^2}{2!} + \ldots \tag{4.15}$$

Since Feynman is determining the relative contribution of $\psi_R(0)$ to $\psi_L(P)$, he is, in effect, determining $\psi_L(P)/\psi_R(0)$, or its equivalent $\psi_R(0) = 1$. The above counting procedure does, therefore, determine a power series in M and N, which does not contain $\psi_R(0)$. Feynman takes the factor $i\varepsilon$ common to all terms out of brackets and uses $M\varepsilon = b$ and $N\varepsilon = a$, which are valid in the continuum limit. Feynman then recognizes the series as the definition of the zeroth-order Bessel function, $J_0\left(2\sqrt{ab}\right)$, or $J_0\left(\sqrt{t^2 - x^2}\cdot\mu\right)$ in the original space and time variables (μ is the mass of the particle).

Using the same method of "path counting", Feynman finds that, for the amplitude to arrive at P moving to the *right*,

$$\psi_R(P) = -\varepsilon\sqrt{\frac{a}{b}}J_1\left(2\sqrt{ab}\right) + \varepsilon\delta(b). \tag{4.16}$$

The last term, $\varepsilon\delta(b)$, "comes formally from [the] derivative of J_0 at $b = 0$, physically from no turn path=" (number 8). If the derivation is not performed, we obtain ($P = (a, b)$):

[4] Cf. Jacobson and Schulman (1984), Figure 1.

$$\psi_R(P) = \varepsilon \frac{\partial}{\partial b} J_0\left(2\sqrt{ab}\right) = -i \frac{\partial}{\partial b} \psi_L(P). \tag{4.17}$$

Therefore,

$$\frac{\partial \psi_L}{\partial b} = i\psi_R \tag{4.18}$$

"is automatic" (Fig. 4.2, number 9). That is, one component (4.12) of the Dirac equation is satisfied. The only thing that still needs to be checked is the other component of the Dirac equation (4.11), that is, one needs to determine $\frac{\partial \psi_R}{\partial a}$, which also comes out "as required": $i\psi_L$ (Fig. 4.2, number 10).

The $\psi_R(a, b)$ and $\psi_L(a, b)$ thus defined are particular solutions to the Dirac equation characterized by their initial values, that is, their values on those lattice points that are the same number of lattice steps away from P as the origin. These are the points on the same horizontal row in the lattice that can be reached by a zigzagging or a straight line going left into $P = (a, b)$. In Fig. 4.3, for instance, the origin is $(a - 2\varepsilon, b - 2\varepsilon)$, that is, it is four lattice steps away from P. Therefore, to give an initial condition for the wave function in this case one needs to specify the value of ψ_L and ψ_R at points $(a - 3\varepsilon, b - \varepsilon)$, $(a - 2\varepsilon, b - 2\varepsilon)$ (the origin), $(a - \varepsilon, b - 3\varepsilon)$ and $(a, b - 4\varepsilon)$.

The boundary condition corresponding to "starting right from zero" is that ψ_L is zero on any of the initial lattice points and ψ_R is 1 at the origin and zero on any other initial lattice point. Such an initial wave function is the discrete version of a wave function describing a point source at the origin. Thus, the solution to Dirac's equation to this particular initial condition is, in the continuum limit, "Green's function" (Fig. 4.2, number 11, and Fig. 4.6, number 1) associated with Dirac's differential equation.

If $\psi_L(P)$ and $\psi_R(P)$ are Green's functions of Dirac's equation, then the wave function at P with more general initial conditions is given by the integral over the initial wave function weighted by the Green's function. The first example Feynman gives is (number 12):

$$\psi_L(x_2, t_2) = i \int J_0\left(\sqrt{(t_2 - t_1)^2 - (x_2 - x_1)^2}\right) \psi_R(x_1, t_1) \frac{dx_1}{2}. \tag{4.19}$$

Here μ was set to 1 for the sake of simplicity and ψ_L was assumed to be zero for any initial point, while "the factor ε just becomes $\frac{dx}{2}$ on integrating Green's function" (number 11).

For arbitrary initial conditions, $\psi_L(P)$ and $\psi_R(P)$ are given by (see Fig. 4.5)

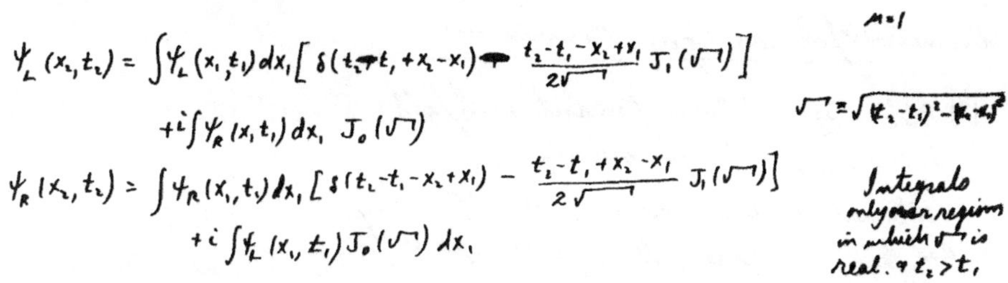

Fig. 4.5 The general solution to Dirac's equation in one dimension, which is obtained by "path counting", ca. 1946 (*Dirac Equation b*, folio 11, reprinted with permission of Melanie Jackson Agency, LLC)

$$\psi_L(x_2,t_2) = \int \psi_L(x_1,t_1)\,dx_1 \left[\delta(t_2-t_1+x_2-x_1) - \frac{t_2-t_1-x_2+x_1}{2\surd}J_1\left(\surd\right)\right]$$
$$+\,i\int \psi_R(x_1t_1)\,dx_1\,J_0\left(\surd\right)$$
$$\psi_R(x_2,t_2) = \int \psi_R(x_1,t_1)\,dx_1 \left[\delta(t_2-t_1-x_2+x_1) - \frac{t_2-t_1+x_2-x_1}{2\surd}J_1\left(\surd\right)\right]$$
$$+\,i\int \psi_L(x_1t_1)\,J_0\left(\surd\right)\,dx_1,$$

where $\surd \equiv \sqrt{(t_2-t_1)^2-(x_2-x_1)^2}$.

4.1.4 Explaining the Wave Function's Time Evolution

As in his thesis and in RMP48, Feynman thus succeeds in defining the wave function at a certain coordinate by the wave function at an earlier coordinate as well as defining a quantity, characteristic of the physical situation at hand, that determines the evolution of the wave function from one coordinate to the other. In both texts, the latter quantity was the integral of the exponential of the classical action that characterizes the physical system along each possible path.

In RMP48, Feynman had also tried to obtain an appropriate action function for the (relativistic) Dirac equation by modifying the action functions corresponding to the Pauli and Klein–Gordon equations. Although the action thus obtained was unjustified, it yielded, however, the Dirac equation. The construction of the action was in that sense *ad hoc* and did not satisfy Feynman, who considered it to be "purely formal" (see Section 3.5).

Now that Feynman has obtained the quantity that determines the evolution of the wave function in a more satisfactory manner, he can interpret Dirac's differential equation in one dimension as the description of the accepted model of the zigzagging electron. The contribution of each path is given by its number of corners (see Fig. 4.6, number 2):

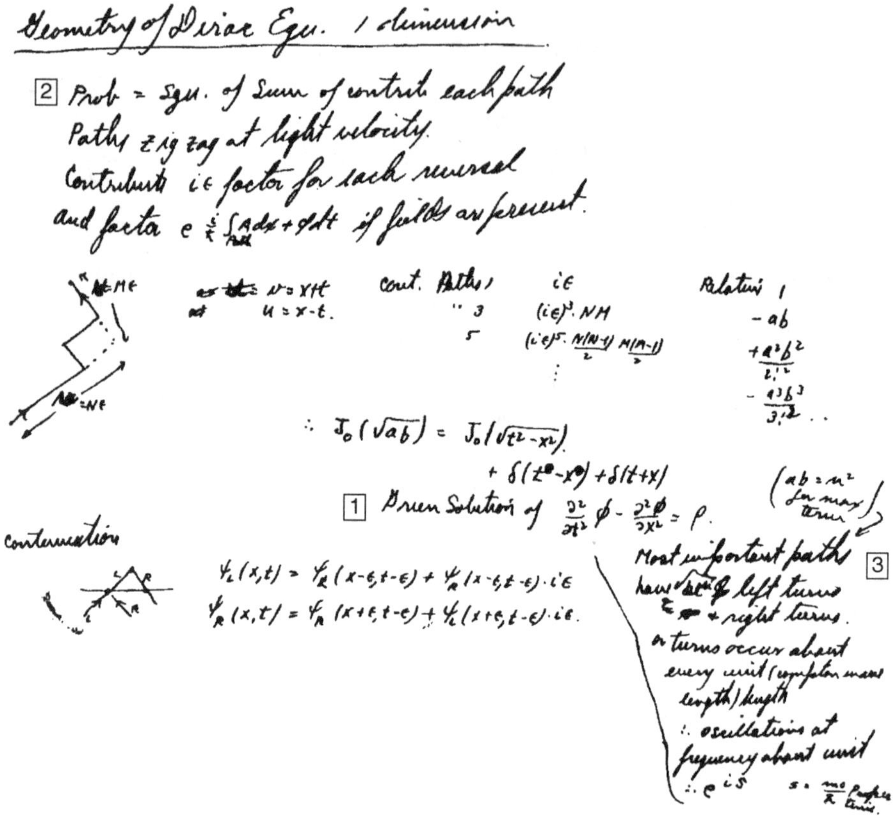

Fig. 4.6 Summary of Feynman's results for the Dirac equation in one spatial dimension, ca. 1946 (*Harmonic Oscillators b*, folio 3, reprinted with permission of Melanie Jackson Agency, LLC)

Prob = Squ. of Sum of contrib. each path

Paths zigzag at light velocity.

Contribute $i\varepsilon$ factor for each reversal

This new way of describing the zigzagging electron by $i\varepsilon$ factors for each change in direction could be used to define the action in this case. Since the exponential of the action, e^{iS}, involved in Feynman's alternative formulation of quantum mechanics, is a complex number of constant absolute value (a unit in fact), it can be read as describing oscillations (number 3):

Most important paths have $\sqrt{\mu}$[?] left turns & right turns. Or turns occur about every unit (compton wave length) length

∴ oscillations at frequency about unit

∴ e^{iS}, $S = \frac{mc}{\hbar}$proper time.

Through the model of an electron zigzagging through an infinitesimally fine space-time lattice, Feynman can now *explain* the time evolution of a relativistic electron, though only in one dimension. And, unlike in the final section of RMP48, Feynman can now justify the action function, since he has derived it from a description of the zigzagging electron.

Also new is the fact that Feynman explicitly refers to the quantity determining the evolution of the wave function as "Green's function" (see Fig. 4.2, number 11; cf. Fig. 4.6, number 1). The use of this new term is evidence of the exposure Feynman had to this type of problem, for instance while working on neutron diffusion during the Second World War.[5]

4.2 Welton's Input: Three Dimensions?

After having successfully treated the one-dimensional Dirac equation, the next step that Feynman was to consider was the Dirac equation describing real electrons, that is, electrons moving not just in one spatial dimension but in three spatial dimensions. Feynman must have communicated his findings concerning the one-dimensional case to his student friend Theodore Welton, who writes, in a letter to Feynman, that he has taken up Feynman's line of thought:

> I was thinking a little about this attempt of yours to understand relativistic Quant. Mech.[6]

Welton had succeeded in treating the Klein–Gordon equation in three dimensions in a way very similar to Feynman's treatment of the one-dimensional Dirac equation (see Fig. 4.7):

> Suppose you make this rule. Given a wave-function $\Psi(\mathbf{x}', t')$ at some time t' for all values of $\mathbf{x}'$. We obtain $\Psi(\mathbf{x}, t)$ at a later time from the recipe that we consider all paths from $(\mathbf{x}', t')$ to $(\mathbf{x}, t)$ along which propagation is at light-velocity at all points. We do not introduce any phase shift at a change in direction, but keep track only of the number of paths and their lengths. The problem is clearly best formulated as an integral equation. We equate $\Psi(\mathbf{x}, t)$ to an alternative expression obtained for it by following back along every possible light path to the last 'scattering':
>
> $$\Psi(\mathbf{x}, t) = \int d\mathbf{x}' \int dt' \frac{-\kappa^2}{4\pi} \delta(s^2) \Psi(\mathbf{x}', t')$$
>
> κ is a length to make the two sides have the same dimensions. If it is desired to consider only forward going light paths, $\delta(s^2)$ can be replaced by a function with twice the singularity on half the light cone & none on the other half. The above equation is obviously just the Klein–Gordon equation:
>
> $$\nabla^2 \Psi - \frac{1}{c^2} \frac{\partial^2 \Psi}{\partial t^2} = \kappa^2 \Psi.$$

By modifying the treatment of the Klein–Gordon equation, Welton is investigating a possible generalization of Feynman's formulation of the Dirac equation to three dimensions (see Fig. 4.8):

---

[5] See, e. g., Galison (1998).

[6] *Dirac Equation a*, folio 27.

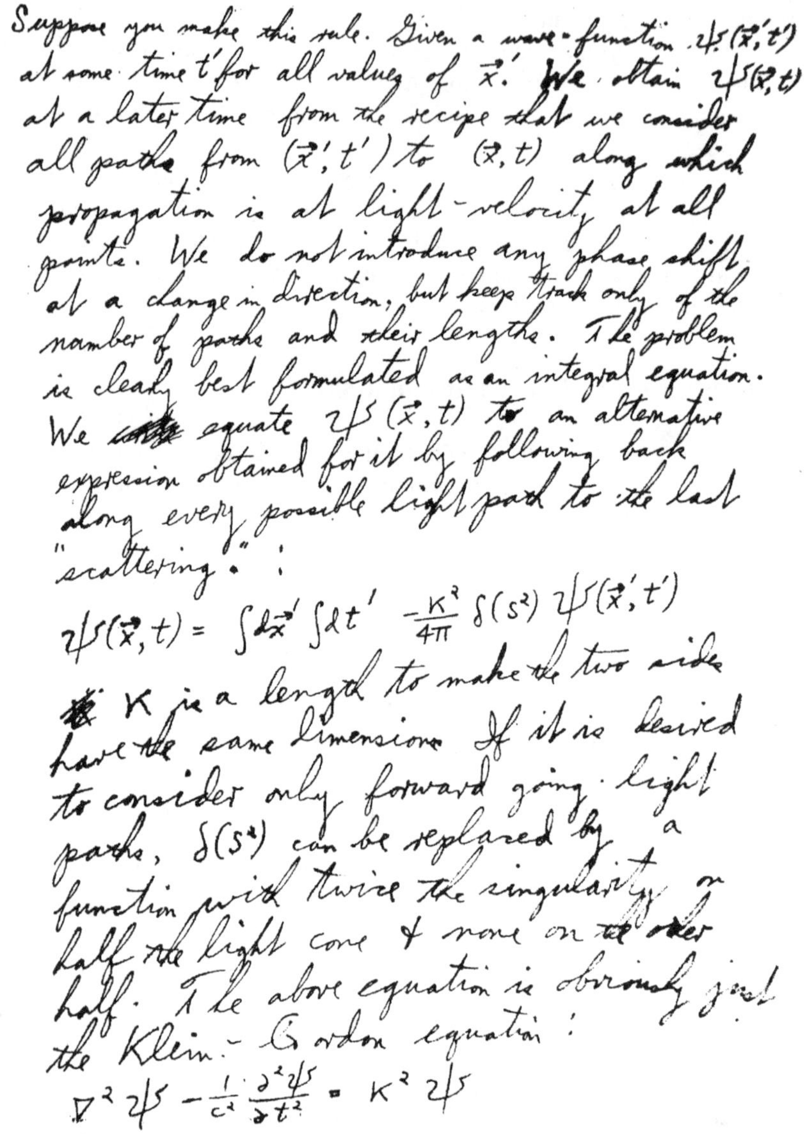

Suppose you make this rule. Given a wave-function $\psi(\vec{x}',t')$ at some time t' for all values of $\vec{x}'$. We obtain $\psi(\vec{x},t)$ at a later time from the recipe that we consider all paths from $(\vec{x}',t')$ to $(\vec{x},t)$ along which propagation is at light-velocity at all points. We do not introduce any phase shift at a change in direction, but keep track only of the number of paths and their lengths. The problem is clearly best formulated as an integral equation. We equate $\psi(\vec{x},t)$ to an alternative expression obtained for it by following back along every possible light path to the last "scattering":

$$\psi(\vec{x},t) = \int d\vec{x}' \int dt' \; -\frac{K^2}{4\pi}\,\delta(s^2)\,\psi(\vec{x}',t')$$

K is a length to make the two sides have the same dimensions. If it is desired to consider only forward going light paths, $\delta(s^2)$ can be replaced by a function with twice the singularity on half the light cone & none on the other half. The above equation is obviously just the Klein-Gordon equation:

$$\nabla^2\psi - \frac{1}{c^2}\frac{\partial^2\psi}{\partial t^2} = K^2\psi$$

Fig. 4.7 Third to last page of the letter that Theodore Welton wrote to Richard Feynman between the end of 1946 and the beginning of 1947 (*Dirac Equation a*, folio 28, reprinted with permission of Melanie Jackson Agency, LLC)

Now according to what you were telling me, the Dirac equation is to be obtained by introducing phase shifts upon each "deflection" of a "light path": I believe you mentioned a factor i for each reversal of direction. I would like to propose a continuous scheme by taking a phase shift something like $e^{\frac{i\theta}{2}}$ where θ is the "deflection angle" of the light, so that no deflection gives a factor 1. I can't give the complete formalism here, but we need a generalization of the $\delta(s^2)$ function which will contain an angular dependence somehow.

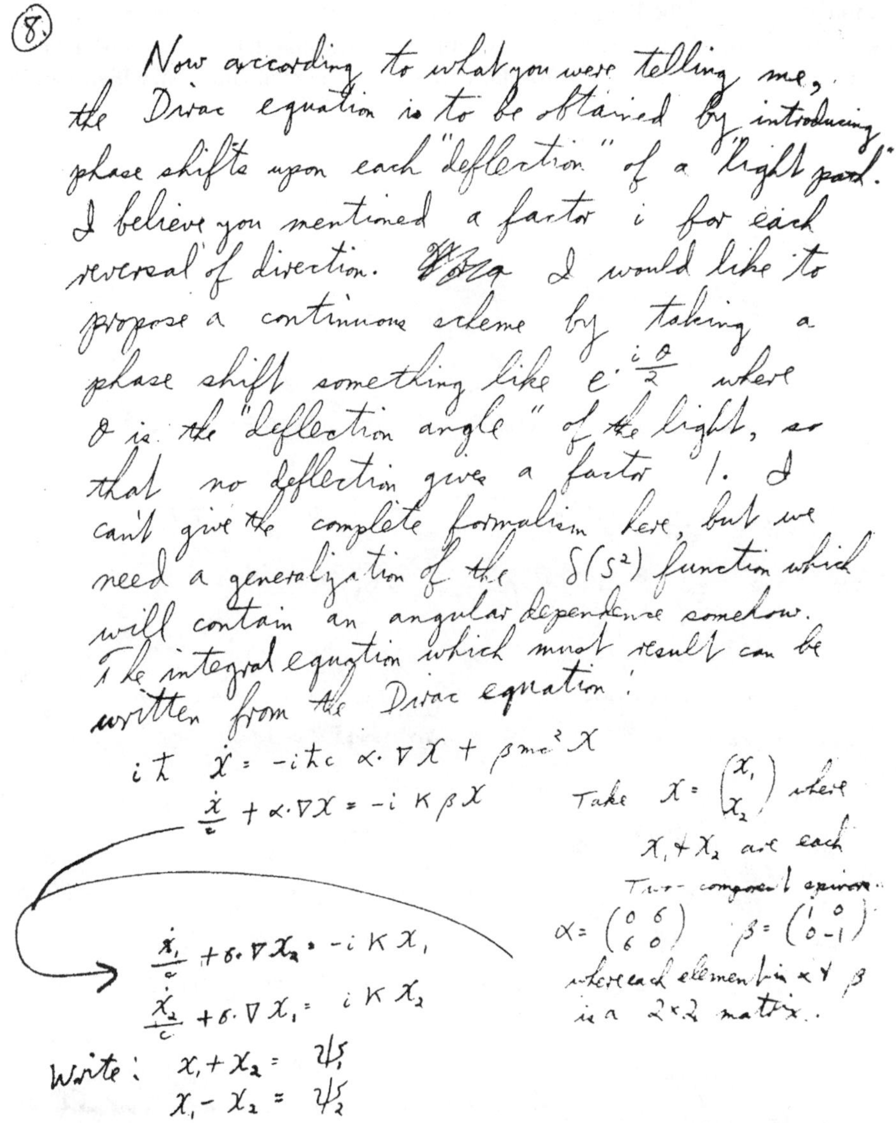

(8) Now according to what you were telling me, the Dirac equation is to be obtained by introducing phase shifts upon each "deflection" of a "light path". I believe you mentioned a factor i for each reversal of direction. I would like to propose a continuous scheme by taking a phase shift something like $e^{\frac{i\theta}{2}}$ where θ is the "deflection angle" of the light, so that no deflection gives a factor 1. I can't give the complete formalism here, but we need a generalization of the $\delta(s^2)$ function which will contain an angular dependence somehow. The integral equation which must result can be written from the Dirac equation:

$$i\hbar\,\dot{\chi} = -i\hbar c\,\alpha\cdot\nabla\chi + \beta mc^2\chi$$

$$\frac{\dot{\chi}}{c} + \alpha\cdot\nabla\chi = -iK\beta\chi$$

Take $\chi = \begin{pmatrix}\chi_1\\ \chi_2\end{pmatrix}$ where $\chi_1 + \chi_2$ are each two-component spinors. $\alpha = \begin{pmatrix}0 & \sigma\\ \sigma & 0\end{pmatrix}$ $\beta = \begin{pmatrix}1 & 0\\ 0 & -1\end{pmatrix}$ where each element in $\alpha + \beta$ is a 2×2 matrix.

$$\frac{\dot{\chi}_1}{c} + \sigma\cdot\nabla\chi_2 = -iK\chi_1$$

$$\frac{\dot{\chi}_2}{c} + \sigma\cdot\nabla\chi_1 = iK\chi_2$$

Write: $\chi_1 + \chi_2 = 2\psi_1$

$\chi_1 - \chi_2 = 2\psi_2$

Fig. 4.8 Penultimate page of the letter that Theodore Welton wrote to Richard Feynman between the end of 1946 and the beginning of 1947 (*Dirac Equation a*, folio 29, reprinted with permission of Melanie Jackson Agency, LLC)

Welton sees that Feynman's alternative formulation basically consists of providing an integral equation rather than a differential equation. It is by using this equation that one can determine the wave function at one coordinate by an integral over the quantity characterizing the paths, which takes the role of a Green's function, and the wave function at another (earlier) coordinate. Welton, therefore, looks for the integral form of the Dirac equation. Although Welton does not succeed in solving the problem, which is significantly more complex than the Klein–Gordon case, completely, he does arrive at a formal expression, the evaluation of which would yield the desired Green's function (see Figs. 4.8 and 4.9):

The integral equation that must result can be written from the Dirac equation:

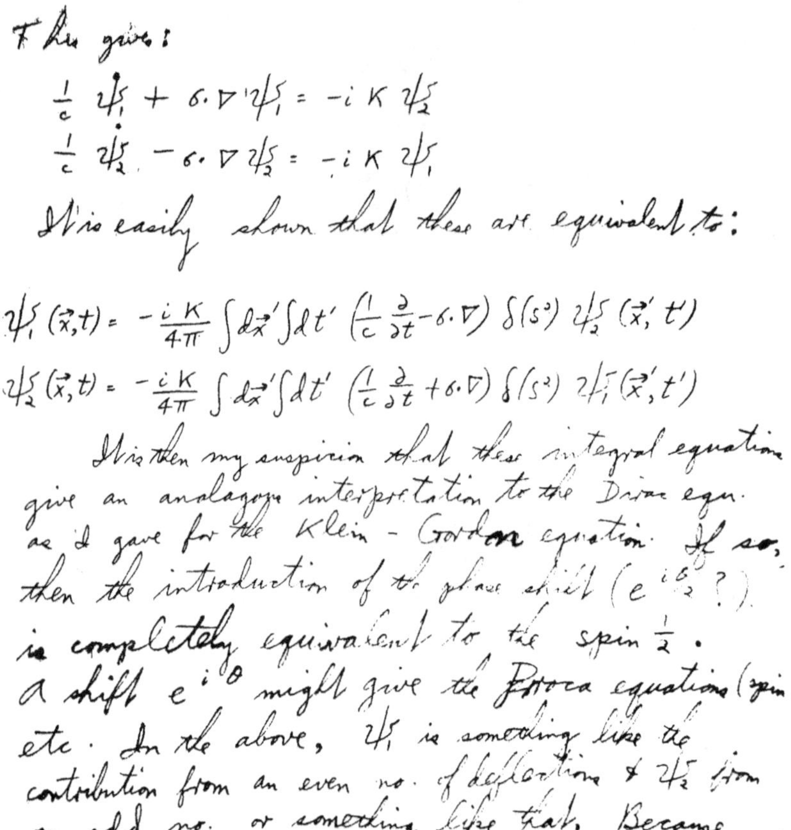
These give:

$$\frac{1}{c}\dot{\psi}_1^s + \sigma\cdot\nabla\psi_1^s = -i\kappa\,\psi_2^s$$

$$\frac{1}{c}\dot{\psi}_2^s - \sigma\cdot\nabla\psi_2^s = -i\kappa\,\psi_1^s$$

It is easily shown that these are equivalent to:

$$\psi_1^s(\vec{x},t) = -\frac{i\kappa}{4\pi}\int d\vec{x}'\int dt'\left(\frac{1}{c}\frac{\partial}{\partial t} - \sigma\cdot\nabla\right)\delta(s^2)\,\psi_2^s(\vec{x}',t')$$

$$\psi_2^s(\vec{x},t) = -\frac{i\kappa}{4\pi}\int d\vec{x}'\int dt'\left(\frac{1}{c}\frac{\partial}{\partial t} + \sigma\cdot\nabla\right)\delta(s^2)\,\psi_1^s(\vec{x}',t')$$

It is then my suspicion that these integral equations give an analogous interpretation to the Dirac eqn. as I gave for the Klein-Gordon equation. If so, then the introduction of the phase shift ($e^{i\sigma/2}$?) is completely equivalent to the spin $\frac{1}{2}$. A shift $e^{i\theta}$ might give the Proca equations (spin 1) etc. In the above, ψ_1^s is something like the contribution from an even no. of deflections & ψ_2^s from an odd no. or something like that. Because of the spinor character of ψ_1 & ψ_2, they have built into them a simple angular dependence which should suffice to fit in with your idea. Boy, is this vague!

Regards,
Ted.

Fig. 4.9 Last page of the letter that Theodore Welton wrote to Richard Feynman between the end of 1946 and the beginning of 1947 (*Dirac Equation a*, folio 30, reprinted with permission of Melanie Jackson Agency, LLC)

$$i\hbar\dot{\chi} = -i\hbar c\alpha \cdot \nabla\chi + \beta mc^2\chi$$

$$\frac{\dot{\chi}}{c} + \alpha \cdot \nabla\chi = -i\kappa\beta\chi$$

Take $\chi = \binom{\chi_1}{\chi_2}$, where $\chi_1 + \chi_2$ are each two component spinors. $\alpha = \begin{pmatrix} 0 & \sigma \\ \sigma & 0 \end{pmatrix}$ $\beta = \begin{pmatrix} 1 & 0 \\ 0 & -1 \end{pmatrix}$, where each element in α & β is a 2×2 matrix.

$$\frac{\dot{\chi}_1}{c} + \sigma \cdot \nabla\chi_2 = -i\kappa\chi_1$$

$$\frac{\dot{\chi}_2}{c} + \sigma \cdot \nabla\chi_1 = i\kappa\chi_2$$

Write:

$$\chi_1 + \chi_2 = \psi_1$$

$$\chi_1 - \chi_2 = \psi_2$$

This gives:

$$\frac{1}{c}\dot{\psi}_1 + \sigma \cdot \nabla\psi_1 = -i\kappa\psi_2$$

$$\frac{1}{c}\dot{\psi}_2 - \sigma \cdot \nabla\psi_2 = -i\kappa\psi_1$$

It is easily shown that these are equivalent to:

$$\psi_1(\mathbf{x}, t) = -\frac{i\kappa}{4\pi}\int d\mathbf{x}'\int dt'\left(\frac{1}{c}\frac{\partial}{\partial t} - \sigma \cdot \nabla\right)\delta(s^2)\psi_2(\mathbf{x}', t')$$

$$\psi_2(\mathbf{x}, t) = -\frac{i\kappa}{4\pi}\int d\mathbf{x}'\int dt'\left(\frac{1}{c}\frac{\partial}{\partial t} + \sigma \cdot \nabla\right)\delta(s^2)\psi_1(\mathbf{x}', t')$$

Once Welton has obtained the integral form of Dirac's equation, he can go on to interpret it in the same way that he interpreted the integral form of the Klein–Gordon equation (Fig. 4.9):

> It is then my suspicion that these integral equations give an analogous interpretation to the Dirac eqn. as I gave for the Klein–Gordon equation.

Welton ends his letter by suggesting a method of interpreting the spinors ψ_1 and ψ_2, which is reminiscent of Feynman's result (see Section 4.1.3) that, from a particle starting on the "right", the "left" component at a later coordinate only receives contributions from paths with an odd number of turns, and the "right" component only receives contributions from paths with an even number of turns. The proposed generalization, however, still needs to be worked out more precisely (Fig. 4.9):

> In the above, Ψ_1 is something like the contribution from an even no. of deflections & Ψ_2 from an odd no. or something like that. Because of the spinor character of Ψ_1 & Ψ_2, they have built into them a simple angular dependence which should suffice to fit in with your idea. Boy, is this vague!

4.3 The Need for an Appropriate Symbol System

Feynman responds to Welton's contribution in a letter by announcing a private research programme:

> I am engaged now in a general program of study—I want to understand (not just in a mathematical way) the ideas in all branches of theor. physics. As you know I am now struggling with the Dirac Eqn.[7]

Despite their vagueness, Feynman recognizes that Welton's ideas might very well "fit in" with his own: "I was especially impressed by your intuition in guessing just what the ψs would mean (I mean the odd & even deflections, etc.)."[8] Indeed, Welton's ideas seem to be so similar that Feynman almost appears anxious about whom should receive the credit for them:

> The ideas which you outlined were, as you say, vague, but otherwise they are exactly right in the sense that the way I now see the Dirac Eqn. is just a more precise formulation of your (& of course my earlier) ideas.[9]

In his reply to Welton, Feynman takes up their discussion of how, in alternative ways, to treat the Dirac equation. In particular Feynman elaborates on the problem of generalizing the promising results found in the one-dimensional case to three spatial dimensions and one time dimension. Welton's line of attack (see Section 4.2) is to adapt the three-dimensional result he obtains for the Klein–Gordon equation to the Dirac equation. Feynman, however, sticks to Dirac's equation and tries to modify the model of the quivering electron and the quantitative description of the electron's change in direction. The fact that Feynman's and Welton's preliminary results and ideas match might be promising, but Feynman recognizes that the cases where more than one spatial dimension are involved will turn out to be particularly complex:

> There are several ways I can explain it, all leading to the same answer but all not equally precise. The one with the most shaky physical ideas at present, I think is closest to the truth & I would like to explain it to you because (1) you may get nice ideas (2) I will learn

[7] *Dirac Equation a* (1946/47a, folio 1), (Feynman's page numbering: 1), see Fig. 4.10; cf. Schweber (1994a, p. 406). Feynman's letter was written on a "Monday February 10". Around the time in question, February 10 was a Monday in 1941, 1947 and 1958. The content of the letter makes 1947 the most plausible date. This is also the date that Schweber assumes to be correct.

[8] *Dirac Equation a* (1946/47a, folio 11 (page 10)), see Fig. 4.15.

[9] *Dirac Equation a* (folio 1 (page 1)).

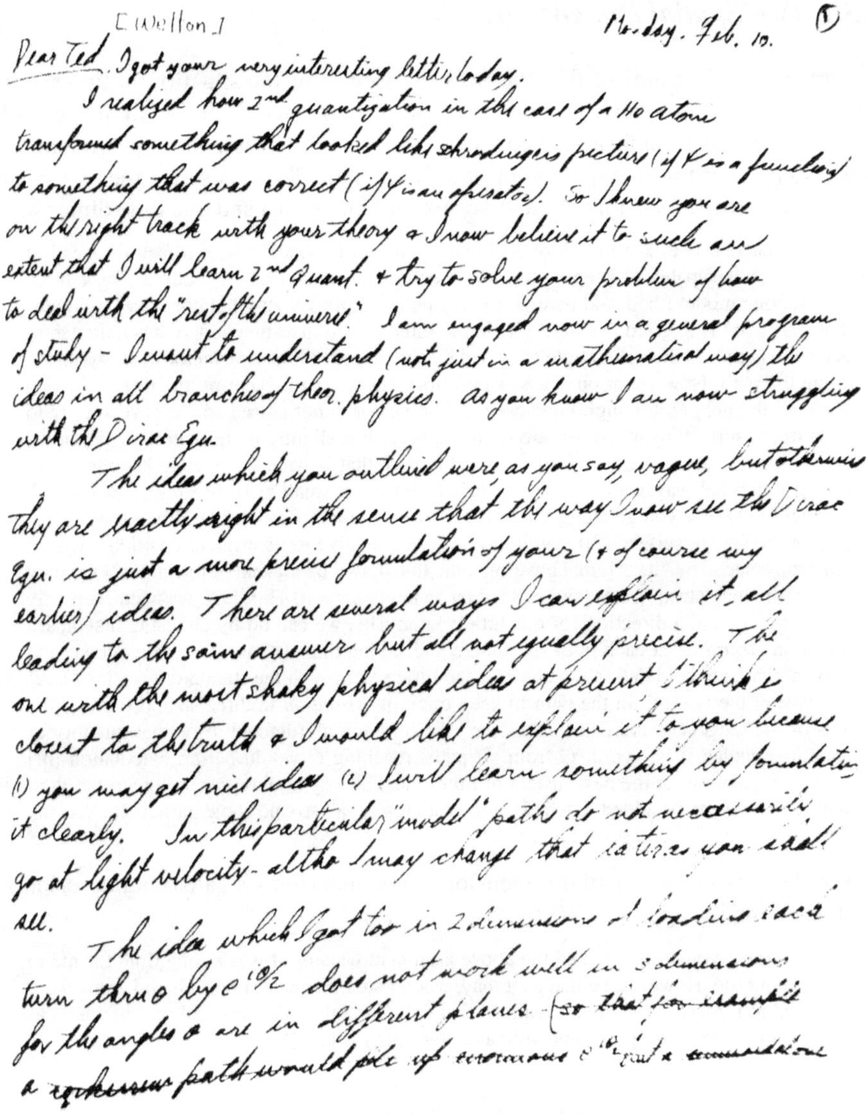
[Welton.] Monday. Feb. 10. (1)

Dear Ted. I got your very interesting letter today.
I realized how 2nd quantization in the case of a He atom transformed something that looked like Schrodinger's picture (if ψ is a function) to something that was correct (if ψ is an operator). So I knew you are on the right track with your theory & I now believe it to such an extent that I will learn 2nd quant. & try to solve your problem of how to deal with the "rest of the universe." I am engaged now in a general program of study - I want to understand (not just in a mathematical way) the ideas in all branches of theor. physics. As you know I am now struggling with the Dirac Eqn.
The ideas which you outlined were, as you say, vague, but otherwise they are exactly right in the sense that the way I now see the Dirac Eqn. is just a more precise formulation of your (& of course my earlier) ideas. There are several ways I can explain it, all leading to the same answer but all not equally precise. The one with the most shaky physical ideas at present I think is closest to the truth & I would like to explain it to you because (1) you may get nice ideas (2) I will learn something by formulating it clearly. In this particular "model" paths do not necessarily go at light velocity - altho I may change that later as you shall see.
The idea which I got too in 2 dimensions of loading each turn thru θ by e^{iθ/2} does not work well in 3 dimensions for the angles θ are in different places - (so that [illegible] a [illegible] path would pile up enormous e^{iθ/2} [illegible]

Fig. 4.10 First page of Feynman's response to Welton's letter, "Monday, Feb. 10. (1947)" *Dirac Equation a* (1946/47a, folio 1, reprinted with permission of Melanie Jackson Agency, LLC)

something by formulating it clearly. In this particular "model" paths do not necessarily go at light velocity—altho I may change that later as you shall see.

The idea which I got too in 2 dimension, of loading each turn thru θ by $e^{i\theta/2}$ does not work well in 3 dimensions for the angles θ are in different places [...][10]

[10] *Dirac Equation a* (folio 1 (page 1)).

4.3.1 Two Spatial Dimensions

Feynman uses diagrams to find the relevant aspects of the situation in two spatial dimensions. He introduces auxiliary elements (arrows and a sphere) in his geometrical descriptions to indicate the characteristic quantities describing the system. He thus tries to determine a more appropriate factor than $e^{i\theta/2}$ to describe the particularities of the three-dimensional case: two spatial dimensions and one time dimension:

> What is needed is to keep track of 3 dimensional rotations, and to use a factor instead of $e^{i\theta/2}$ which is capable of keeping track of a complete 3-dimensional rotation. To be more precise, one model I had said this: go on any path at light velocity but consider only paths that start from A & go to B entering B in the same direction as they left A. Then the factor is $e^{ia/2}$ where a is the solid angle enclosed by the tangent (velocity) vector in its gyrations along the path. Now we at once ask what information must be kept by the time we reach C. At C the area (solid angle) enclosed by the vector is not closed so we have simply to keep track, sort of, of where we are on the sphere. But slightly more than that. When we get back to A (at B) we have [an] enclosed area & that is what we have to keep track of. One geometrical way we can keep track is to put up a small arrowhead at A & move it parallel to itself as you go along the trajectory (differential parallelism). When it gets to B ([?] on the sphere surface [?]) it will have rotated, relative to its original position [crossed out text replaced by "its original position": the line drawn by the end of the large[?] velocity vector (the indicatrix[?], lets [sic] call it)] by an angle $2\pi - a$. Thus at C we need keep only the axis of C and a direction for our little vector. This we can do by choosing a standard direction arbitrarily at each pt of the sphere and saying that at C we have to rotate around C by a given angle to define the angle of the little vector with the standard direction. Thus we have at the point C in the path to keep track of a *rotation* in three spatial (around the axis of the tangent to the path). Thus the "wave function" (defined to be that quantity or quantities which if known at C from *all* paths reaching C would permit calculation [of] the same quantities at the next instant of time) must be a symbol representing a rotation in space—which can be added & which, if the rotation is "a" around some particular axis (say the z axis) must be represented by $e^{ia/2}$.[11]

For the case of two spatial dimensions, Feynman finds a satisfying description (and, incidentally, appeases Welton):

> I don't expect you to understand the above argument because it was a long time for me to figure it out too. However, be that as it may, when I studied quaternions which I knew were designed to represent rotations I realized that they were the mathematical tool in which to represent my thoughts. To remind you: a quaternion q is [...][12]

4.3.2 Three Spatial Dimensions

In contrast to the case of two spatial dimensions, when it comes to three spatial dimensions, ordinary quaternions are an insufficient tool. Feynman, therefore, has to find another way "to represent [his] thoughts". *Complex* quaternions are one

[11] *Dirac Equation a*, folio 2 (page 2), see Fig. 4.11, emph. in the original.

[12] *Dirac Equation a*, folio 3 (page 3), see Fig. 4.12.

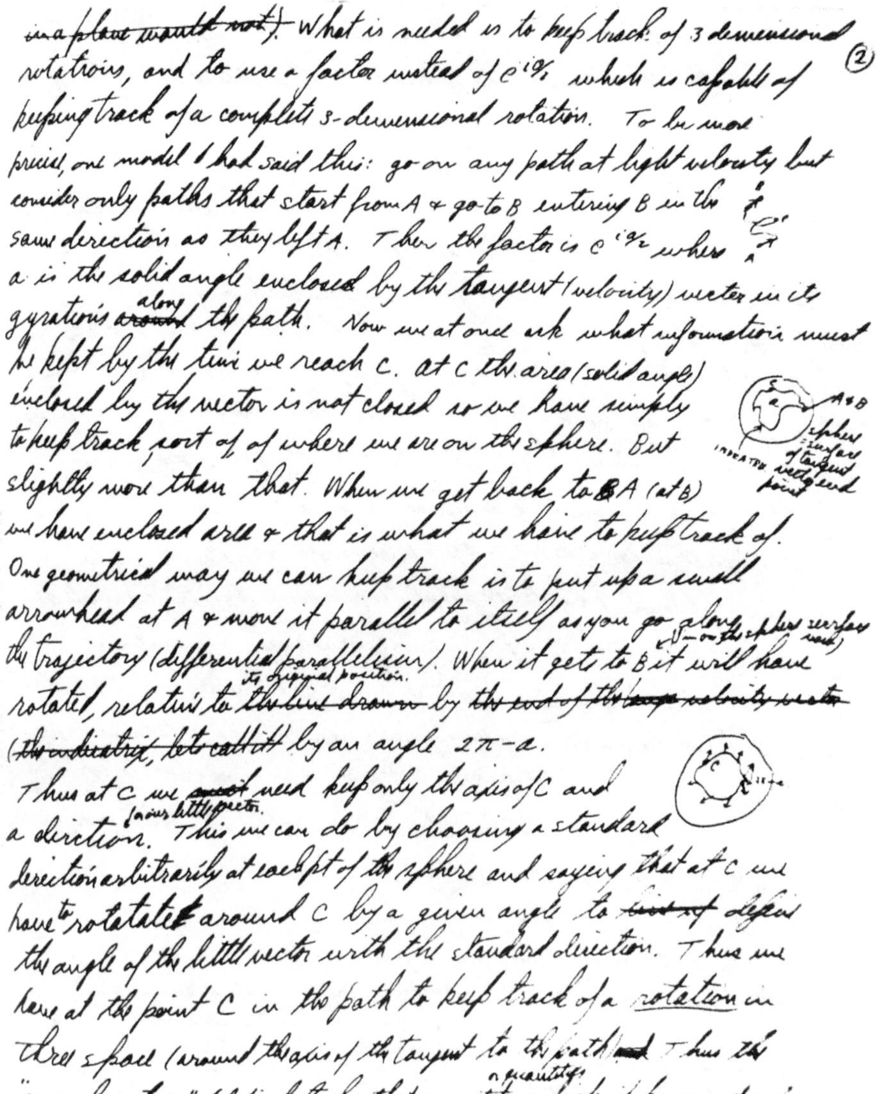

~~in a plane wouldn't not)~~. What is needed is to keep track of 3 dimensional rotations, and to use a factor instead of $e^{i\theta/2}$ which is capable of keeping track of a complete 3-dimensional rotation. To be more precise, one model I had said this: go on any path at light velocity but consider only paths that start from A & go to B entering B in the same direction as they left A. Then the factor is $e^{ia/2}$ where a is the solid angle enclosed by the tangent (velocity) vector in its gyrations ~~around~~ along the path. Now we at once ask what information must be kept by the time we reach C. At C the area (solid angle) enclosed by the vector is not closed so we have simply to keep track, sort of, of where we are on the sphere. But slightly more than that. When we get back to ~~B~~A (at B) we have enclosed area & that is what we have to keep track of. One geometrical way we can keep track is to put up a small arrowhead at A & move it parallel to itself as you go along the trajectory (differential parallelism) (— on the sphere surface). When it gets to B it will have rotated, relative to its original position ~~the line drawn~~ by ~~the end of the tangent velocity vector~~ (~~the indicatrix, let's call it~~) by an angle $2\pi - a$.

Thus at C we ~~must~~ need keep only the axis of C and a direction (a is little vector). This we can do by choosing a standard direction arbitrarily at each pt of the sphere and saying that at C we have to rotate ~~it~~ around C by a given angle to ~~find~~ define the angle of the little vector with the standard direction. Thus we have at the point C in the path to keep track of a rotation in three space (around the axis of the tangent to the path) ~~and~~ Thus the "wave function" (defined to be that a quantity which if known at C from

Fig. 4.11 Second page of Feynman's response to Welton's letter (February 1947): The section of Feynman's letter to Welton in which he expounds the generalization to two spatial dimensions of his treatment of the Dirac equation in one spatial dimension (*Dirac Equation a*, folio 2, reprinted with permission of Melanie Jackson Agency, LLC)

possibility. However, he could use them to treat the case of four equivalent dimensions. But according to the theory of relativity, the time dimension is singled out by a negative sign of its contribution to the metric of space-time. Therefore, complex quaternions do not quite serve Feynman's objective. He again sympathizes with Welton and then describes the situation in which he finds himself:

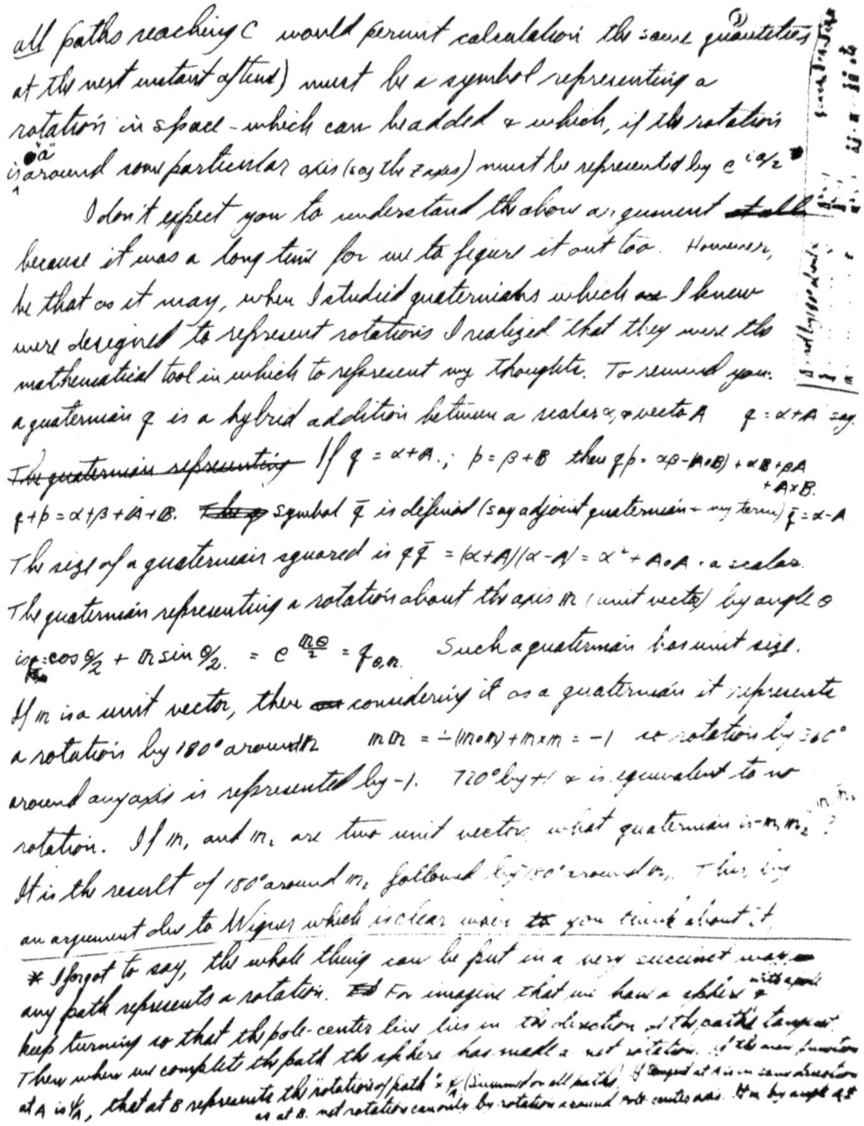

all paths reaching C would permit calculation the same quantities (2) at the next instant of time) must be a symbol representing a rotation in space - which can be added & which, if the rotation is around some particular axis (say the z axis) must be represented by $e^{i\alpha/2}$

I don't expect you to understand the above argument ~~at all~~ because it was a long time for me to figure it out too. However, be that as it may, when I studied quaternions which I knew were designed to represent rotations I realized that they were the mathematical tool in which to represent my thoughts. To remind you: a quaternion q is a hybrid addition between a scalar α & vector A $q = \alpha + A$ say.

~~The quaternion representing~~ If $q = \alpha + A$, $p = \beta + B$ then $qp = \alpha\beta - (A\cdot B) + \alpha B + \beta A + A\times B$.

$q + p = \alpha + \beta + A + B$. ~~The~~ symbol $\bar{q}$ is defined (say adjoint quaternion — my term) $\bar{q} = \alpha - A$

The size of a quaternion squared is $q\bar{q} = (\alpha+A)(\alpha-A) = \alpha^2 + A\cdot A$ = a scalar.

The quaternion representing a rotation about the axis n (unit vector) by angle θ is $\cos\theta/2 + n\sin\theta/2 = e^{n\theta/2} = q_{\theta,n}$. Such a quaternion has unit size.

If n is a unit vector, then considering it as a quaternion it represents a rotation by 180° around n. $nn = -(n\cdot n) + n\times n = -1$ so rotation by 360° around any axis is represented by -1. 720° by $+1$ & is equivalent to no rotation. If n_1 and n_2 are two unit vectors, what quaternion is $-n_1n_2$? It is the result of 180° around n_2 followed by 180° around n_1. This is an argument due to Wigner which is clear when you think about it.

* I forgot to say, the whole thing can be put in a very succinct way: any path represents a rotation. For imagine that we have a sphere with a pole & keep turning so that the pole-center line lies in the direction of the path's tangent. Then when we complete the path the sphere has made a net rotation. If the [illegible] function at A is ψ_A, that at B represents the rotation of path $\times\,\psi_A$ (Summed over all paths) [illegible]

Fig. 4.12 Third page of Feynman's letter to Welton, February 1947 (*Dirac Equation a*, folio 3, reprinted with permission of Melanie Jackson Agency, LLC)

> Now, if you are still staggering but on your feet I will tell you more which is even harder to understand. [...] We desire a symbolism, which by analogy works in 4 (3+1) space like the quaternions do in 3. I shall show that this symbolism is furnished by quaternions using complex numbers for components! I have trouble visualizing 3+1 space, but I have become an expert at visualizing (really!) 4 dimensions with all dimensions equivalent. $ds^2 = dx^2 + dy^2 + dz^2 + [dt^2]$.[13]

[13] *Dirac Equation a*, folios 5 and 6 (pages 5 and 6) see Figs. 4.13 and 4.14.

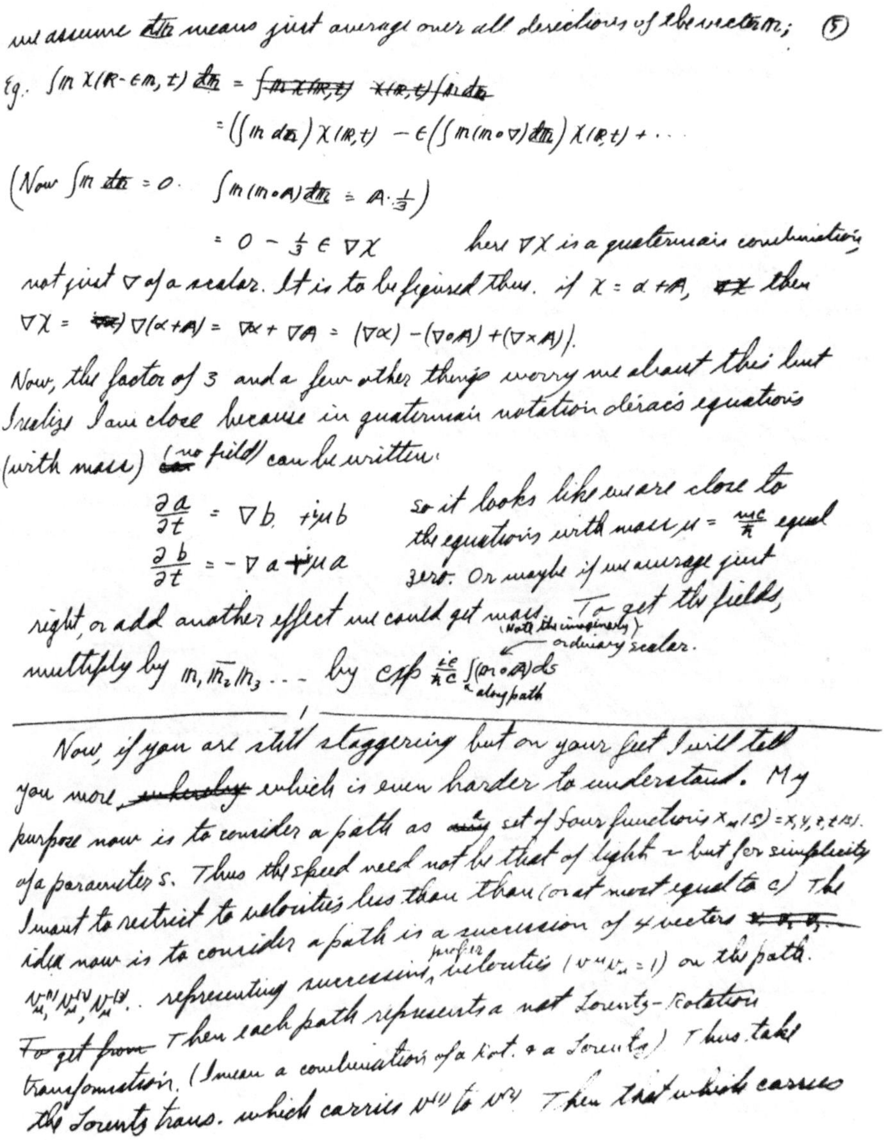
we assume $d\mathbf{m}$ means just average over all directions of the vector $\mathbf{m}$; (5)

Eg. $\int \mathbf{m}\, \chi(\mathbf{R}-\epsilon\mathbf{m}, t)\, d\mathbf{m} =$

$= \left(\int \mathbf{m}\, d\mathbf{m}\right) \chi(\mathbf{R},t) - \epsilon\left(\int \mathbf{m}(\mathbf{m}\circ\nabla) d\mathbf{m}\right)\chi(\mathbf{R},t) + \ldots$

(Now $\int \mathbf{m}\, d\mathbf{m} = 0$. $\int \mathbf{m}(\mathbf{m}\circ \mathbf{A}) d\mathbf{m} = \mathbf{A}\cdot\frac{1}{3}$)

$= 0 - \frac{1}{3}\epsilon \nabla\chi$ here $\nabla\chi$ is a quaternion combination not just ∇ of a scalar. It is to be figured thus. if $\chi = \alpha + \mathbf{A}$, then $\nabla\chi = \nabla(\alpha + \mathbf{A}) = \nabla\alpha + \nabla\mathbf{A} = (\nabla\alpha) - (\nabla\circ\mathbf{A}) + (\nabla\times\mathbf{A})$).

Now, the factor of 3 and a few other things worry me about this but I realize I am close because in quaternion notation dirac's equations (with mass) (no field) can be written:

$\frac{\partial a}{\partial t} = \nabla b + i\mu b$

$\frac{\partial b}{\partial t} = -\nabla a + i\mu a$

so it looks like we are close to the equations with mass, $\mu = \frac{mc}{\hbar}$ equal zero. Or maybe if we average just right, or add another effect we could get mass. (Note the imaginary) To get the fields, multiply by $m_1, m_2, m_3 \ldots$ by $\exp \frac{ie}{\hbar c}\int(\mathbf{m}\circ\mathbf{A})ds$ along path ← ordinary scalar.

Now, if you are still staggering but on your feet I will tell you more, which is even harder to understand. My purpose now is to consider a path as set of four functions $x_\mu(s) = x, y, z, t(s)$. of a parameter s. Thus the speed need not be that of light – but for simplicity I want to restrict to velocities less than than (or at most equal to c) The idea now is to consider a path is a succession of 4 vectors proper velocities ($v^\mu v_\mu = 1$) on the path. $v_\mu^{(1)}, v_\mu^{(2)}, v_\mu^{(3)}$... representing successive, Then each path represents a not Lorentz-Rotation transformation. (I mean a combination of a Rot. + a Lorentz) Thus take the Lorentz trans. which carries $v^{(1)}$ to $v^{(2)}$ Then that which carries

Fig. 4.13 Fifth page of Feynman's letter to Welton, February 1947 (*Dirac Equation a*, folio 5, reprinted with permission of Melanie Jackson Agency, LLC)

By "visualizing" Feynman does not mean here the ability to produce a mental image of an abstract object. Rather, Feynman needs a "mathematical tool in which to represent [his] thoughts". He has to construct an appropriate symbol system through which certain propositions can be expressed, and by means of which, like with an instrument as it were, other symbols corresponding to entailed propositions can be deduced.

$N^{(4)}$ to $N^{(3)}$ etc. What is the net? ~~By~~ We desire a symbolism, which
by analogy works in (3+1) 4 space like the quaternions do in 3. I shall (6)
show that this symbolism is furnished by quaternions ~~and~~ using
complex numbers for ~~ac~~ components!

I have trouble visualizing 3+1 space, but I have become an expert
at visualizing (really!) 4 dimensions with all dimensions equivalent. $ds^2 = dx^2 + dy^2 + dz^2 + dw^2$.
What are the properties of rotations in this space? First a simple
rotation (called turn) from one vector m_1 into ~~and~~ another m_2 is not around just one axis
but around two (since there are 2 perpend. directions both ⊥ to m_1 & m_2).
More precisely for any plane (eg of m_1 & m_2) there is a completely ⊥ plane
(I call it the antipodal plane) such that if m' is any vector in the first plane, & m^2 is
any vector in the second plane, then m' is ⊥ to m^2. ~~A general rota~~
a plane (& hence its antipodal plane) is determined by 4 parameters {2 perpendicular
unit vectors = 8 − 2^{unit} − $1^{\text{perp.}}$ = 5 nos. but these can be set at any angle in the plane ∴ 4 para.}
a complete 4 dim. rot. is characterized by 6 parameters & can always
be resolved into a ~~rotation~~ turn (1 parameter) in a certain plane (4 para.) and in addition
some other rotational angle (turn) in the antipodal plane (1 para.). ~~Rotations~~
Turns in two antipodal planes commute, obviously. (Example a Lorentz trans
from velocity in z direction is a turn in the zt plane, this commutes with space
rotations around the z axis (turns in the xy plane)).

Now (lets call these octonians) the "quaternion" representing 4-space
rotations will probably have among their fundamental unit vectors, a set
of six: $I_{xy}, I_{yz}, I_{zx}, I_{xw}, I_{yw}, I_{zw}$ such that $I_{xy}^2 = -1$ etc. and such relations
as are true with one dimension lacking are still true, such as $I_{xy} I_{yz} = I_{zx} = -I_{yz} I_{xy}$
(We can identify $I_{xy} = k$, $I_{yz} = i$, $I_{zx} = j$). These fundamental units represent
~~rotations~~ turns of 180° in the various planes xy etc. ~~E.~~ Hypothesis; the
Octonian representing a rotation of φ in the xy plane is $e^{\frac{\varphi}{2} I_{xy}} = \cos\frac{\varphi}{2} + \sin\frac{\varphi}{2} I_{xy}$.
Etc. for other planes. II If we go φ on xy plane, and θ on antipodal (zw) plane

Fig. 4.14 Sixth page of Feynman's letter to Welton, February 1947 (*Dirac Equation a*, folio 6, reprinted with permission of Melanie Jackson Agency, LLC)

There follows a long, detailed and complicated explication of his ideas about an alternative formulation of the "four-dimensional" and the "3 + 1"-dimensional Dirac equations. Feynman partially succeeds in achieving his goal of constructing an appropriate symbol system out of the complex quaternions. He understands which relevant aspects of the phenomena some of Dirac's symbols "keep track of" and compares these aspects with those of his alternative symbol system:

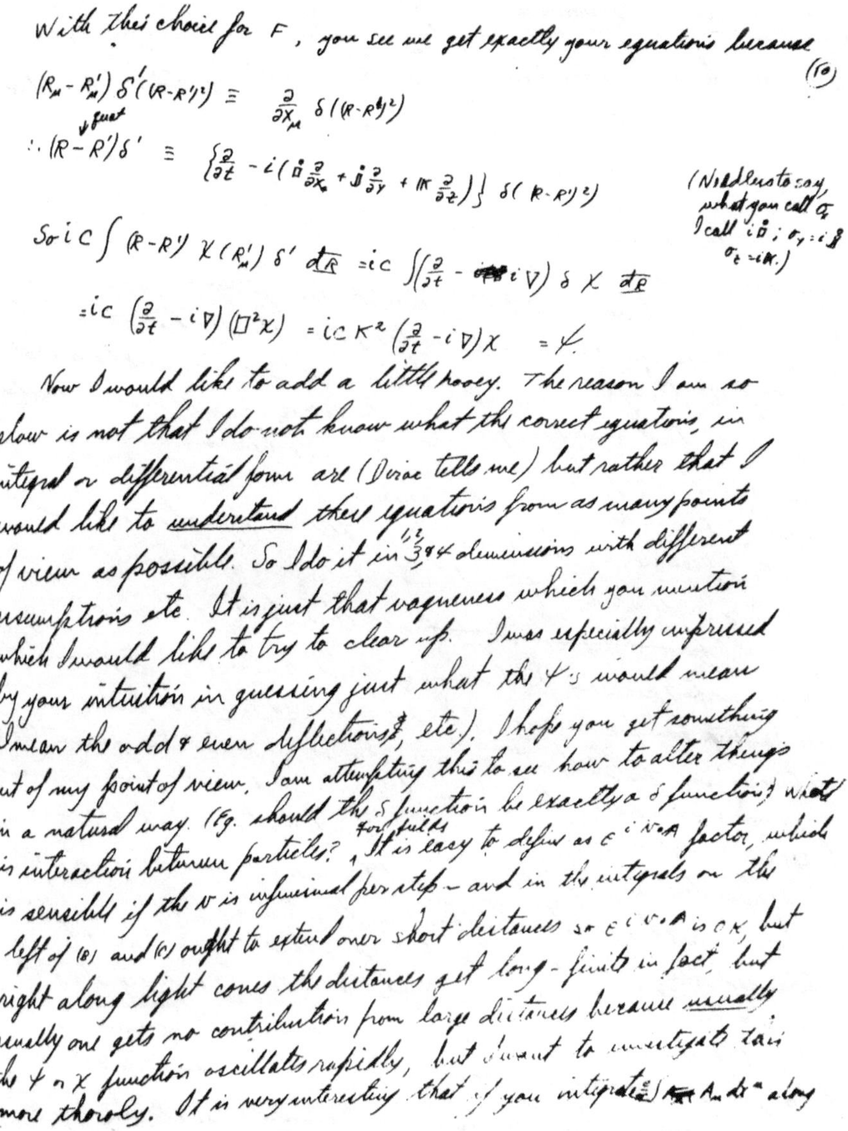
With this choice for F, you see we get exactly your equations because

$(R_\mu - R'_\mu)\,\delta'((R-R')^2) \equiv \frac{\partial}{\partial X_\mu}\,\delta((R-R')^2)$ (10)

just

$\therefore (R-R')\,\delta' \equiv \left\{\frac{\partial}{\partial t} - i\left(\mathbf{i}\frac{\partial}{\partial x} + \mathbf{j}\frac{\partial}{\partial y} + \mathbf{k}\frac{\partial}{\partial z}\right)\right\}\delta((R-R')^2)$

(Needless to say, what you call σ_x I call $i\mathbf{i}$; $\sigma_y = i\mathbf{j}$; $\sigma_z = i\mathbf{k}$.)

So $ic\int (R-R')\,\chi(R'_\mu)\,\delta'\,d\tau_R = ic\int\left(\frac{\partial}{\partial t} - i\nabla\right)\delta\,\chi\,d\tau_R$

$= ic\left(\frac{\partial}{\partial t} - i\nabla\right)(\Box^2\chi) = icK^2\left(\frac{\partial}{\partial t} - i\nabla\right)\chi = \psi.$

Now I would like to add a little hooey. The reason I am so slow is not that I do not know what the correct equations, in integral or differential form are (Dirac tells me) but rather that I would like to understand these equations from as many points of view as possible. So I do it in 1, 2, 3 & 4 dimensions with different assumptions etc. It is just that vagueness which you mention which I would like to try to clear up. I was especially impressed by your intuition in guessing just what the ψ's would mean (I mean the odd & even deflections, etc). I hope you get something out of my point of view. I am attempting this to see how to alter things in a natural way. (Eg. should the δ function be exactly a δ function? what is interaction between particles? for fields, It is easy to define as $e^{i\nu\cdot A}$ factor, which is sensible if the ν is infinitesimal per step – and in the integrals on the left of (8) and (9) ought to extend over short distances so $e^{i\nu\cdot A}$ is OK, but right along light cones the distances get long – finite in fact, but usually one gets no contribution from large distances because usually the ψ or χ function oscillates rapidly, but I want to investigate this more thoroly. It is very interesting that if you integrate $A_\mu dx^\mu$ along

Fig. 4.15 Tenth page of Feynman's letter to Welton, February 1947 (*Dirac Equation a*, folio 11, reprinted with permission of Melanie Jackson Agency, LLC)

> For example, I'm beginning to get a [m?]ild "understanding" of the place of Dirac[']s α matrices which were invented by him "to produce an equation of first order in the differential coefficient in the time," but by me in order "to keep track of the result of a succession of changes of coordinate system."[14]

---

[14] *Dirac Equation a*, folios 12 and 14 (pages 11 and 12), see Figs. 4.16 and 4.17.

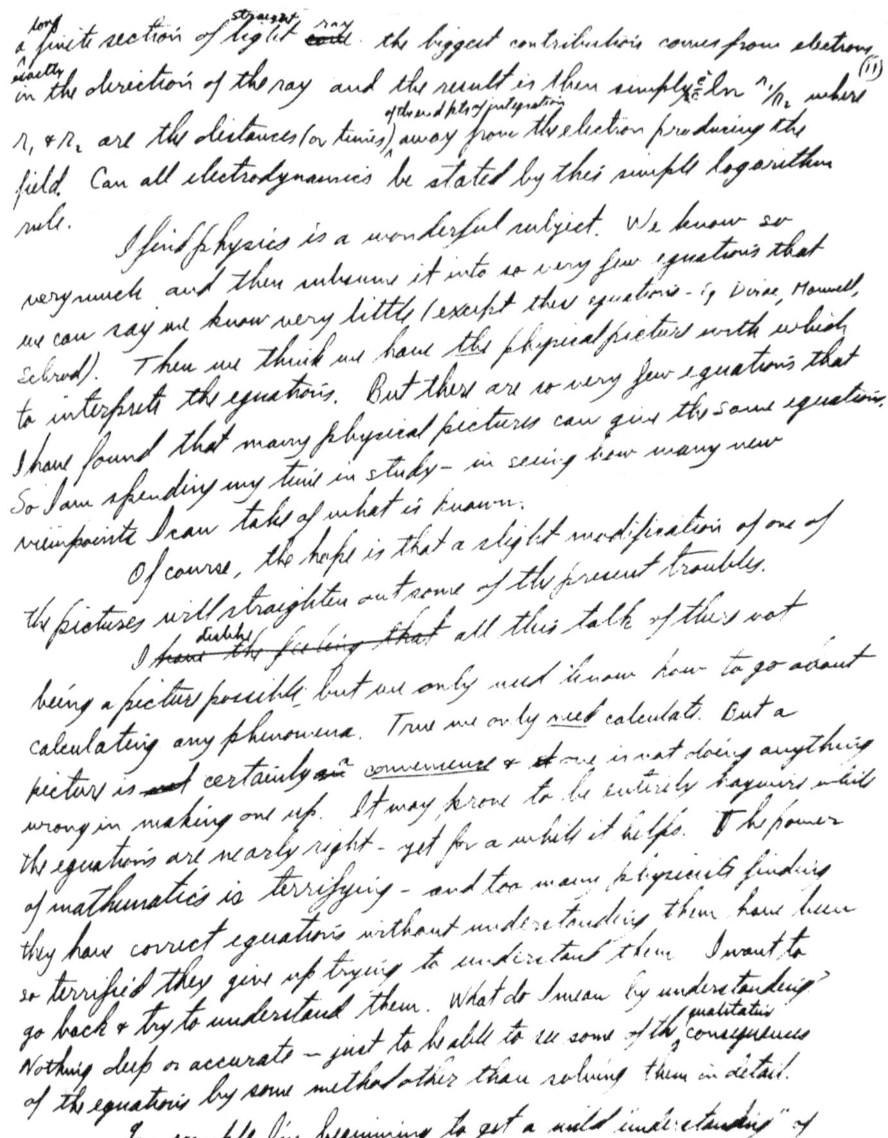
a long finite section of straight light ray the biggest contribution comes from electrons
exactly in the direction of the ray and the result is then simply $\frac{e}{c} \ln r_1/r_2$ where (11)
r_1 & r_2 are the distances (or times) of the end pts of integration away from the electron producing the
field. Can all electrodynamics be stated by this simple logarithm
rule.
I find physics is a wonderful subject. We know so
very much and then subsume it into so very few equations that
we can say we know very little (except these equations - eg Dirac, Maxwell,
Schrod). Then we think we have the physical picture with which
to interpret the equations. But there are so very few equations that
I have found that many physical pictures can give the same equations.
So I am spending my time in study - in seeing how many new
viewpoints I can take of what is known.
Of course, the hope is that a slight modification of one of
the pictures will straighten out some of the present troubles.
I ~~have the feeling that~~ doubt all this talk of there not
being a picture possible but we only need know how to go about
calculating any phenomena. True we only need calculate. But a
picture is ~~not~~ certainly a convenience & one is not doing anything
wrong in making one up. It may prove to be entirely hogwash while
the equations are nearly right - yet for a while it helps. The power
of mathematics is terrifying - and too many physicists finding
they have correct equations without understanding them have been
so terrified they give up trying to understand them. I want to
go back & try to understand them. What do I mean by understanding?
Nothing deep or accurate - just to be able to see some of the qualitative consequences
of the equations by some method other than solving them in detail.
For example, I'm beginning to get a "wild understanding" of

Fig. 4.16 Eleventh page of Feynman's letter to Welton, February 1947 (*Dirac Equation a*, folio 12, reprinted with permission of Melanie Jackson Agency, LLC)

On the whole, however, Feynman is not satisfied with his results. Although the mathematical relations do indeed come out correctly, he cannot physically interpret the relationships. To proceed towards his goal of fully "understanding" the Dirac equation, Feynman plans to re-use his "reformulation" panacea:

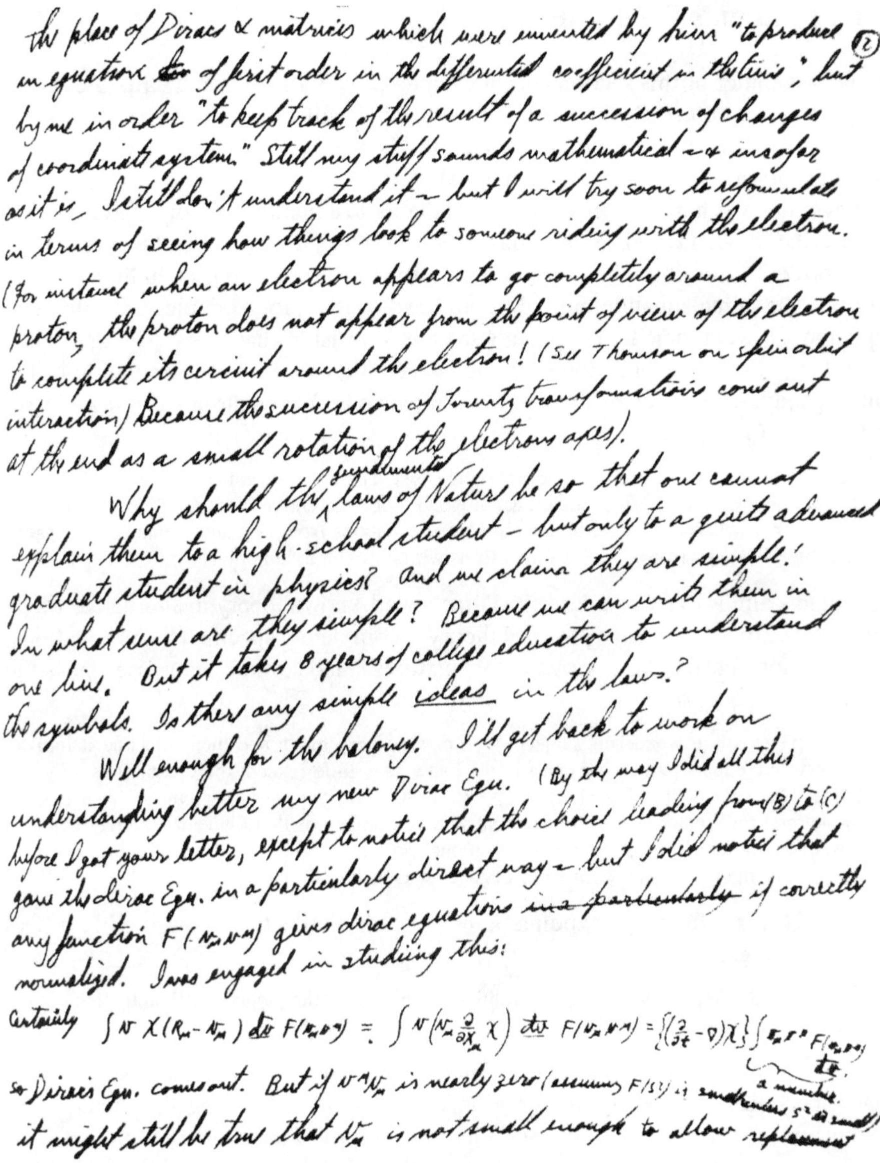
the place of Dirac α matrices which were invented by him "to produce an equation ~~for~~ of first order in the differential coefficient in the time"; but by me in order "to keep track of the result of a succession of changes of coordinate system." Still my stuff sounds mathematical – & insofar as it is, I still don't understand it – but I will try soon to reformulate in terms of seeing how things look to someone riding with the electron. (For instance when an electron appears to go completely around a proton, the proton does not appear from the point of view of the electron to complete its circuit around the electron! (See Thomson on spin orbit interaction) Because the succession of Lorentz transformations come out at the end as a small rotation of the electrons axes). ⑫

Why should the fundamental laws of Nature be so that one cannot explain them to a high-school student – but only to a quite advanced graduate student in physics? And we claim they are simple! In what sense are they simple? Because we can write them in one line. But it takes 8 years of college education to understand the symbols. Is there any simple *ideas* in the laws?

Well enough for the baloney. I'll get back to work on understanding better my new Dirac Eqn. (By the way I did all this before I got your letter, except to notice that the choice leading from (B) to (C) gave the Dirac Eqn. in a particularly direct way – but I did notice that any function F(N, N·M) gives dirac equations ~~in a particularly~~ if correctly normalized. I was engaged in studying this:

Certainly $\int N\, X(R_m - N_m)\, d\Pi\, F(N_m, N\cdot M) = \int N\left(N_m \frac{\partial}{\partial X_m} X\right) d\Pi\, F(N_m, N\cdot M) = \left\{\left(\frac{\partial}{\partial t} - \nabla\right)X\right\}\int \ldots F(N_m, N\cdot M)\, d\Pi$

so Dirac's Eqn. comes out. But if $N^\mu N_\mu$ is nearly zero (assuming F(S) is a number small unless S^2 is small) it might still be true that N_m is not small enough to allow replacement

Fig. 4.17 Twelfth page of Feynman's letter to Welton, February 1947 (*Dirac Equation a*, folio 14, reprinted with permission of Melanie Jackson Agency, LLC)

Still my stuff sounds mathematical—& insofar as it is, I still don't understand it—but I will try soon to reformulate in terms of seeing how things look to someone riding with the electron.[15]

[15] *Dirac Equation a*, folio 14 (page 12).

4.4 Feynman's Programme

In the remainder of this section and in Chapter 5 we will see that the first Feynman diagrams (i. e. before Dyson's intervention that I will discuss in Chapter 6) are a product of Feynman's efforts, announced in his reply to Welton, to understand the Dirac equation "not just in a mathematical way".

Feynman was not searching for new mathematical methods or for a new equation to describe the behaviour of electrons. Few physicists doubted that the Dirac equation correctly described the behaviour of a relativistic electron, including its spin. Nor was the Dirac equation more difficult to solve than, for example, Schrödinger's equation. Of what then did Feynman's struggle consist? What was he trying to find?

In fact, after the explication of his ideas about an alternative formulation of the Dirac equation in different numbers of dimensions, Feynman has given a clearer indication of his goals:

> Now I would like to add a little hooey [nonsense]. The reason I am so slow is not that I do not know what the correct equations, in integral or differential form are (Dirac tells me) but rather that I would like to *understand* these equations from as many points of view as possible. So I do it in 1, 2, 3, & 4 dimensions with different assumptions etc.[16]

Feynman's aim is to try and describe Dirac's well-known theory in alternative ways, for he does not believe that a physical theory is completely specified by its equations. The equations have to be completed by "pictures", and several pictures are possible for the same equations:

> I find physics is a wonderful subject. We know so very much and then subsume it into so very few equations that we can say we know very little (except these equations—Eg. Dirac, Maxwell, Schrod[inger]). Then we think we have *the* physical picture with which to interpret the equations. But there are so very few equations that I have found that many physical pictures can give the same equations. So I am spending my time in study—in seeing how many new viewpoints I can take of what is known.[17]

The search for different viewpoints is not just an intellectual "pleasure"[18]; it also has a precise goal:

> Of course, the hope is that a slight modification of one of the pictures will straighten out some of the present troubles.[19]

Feynman's objective is to be able to interpret the known equations in such a way that it becomes clear which assumption in the theory is causing the inconsistent conclusions in the troublesome cases. Once the culprit of the contradiction (between the theory and more general physical principles or uncontested experimental data) has been identified, it should then be possible to resolve the problem by "modifying" the problematic assumption.

---

[16] *Dirac Equation a*, folio 11 (page 10).

[17] *Dirac Equation a*, folio 12 (page 11).

[18] RMP48, p. 367.

[19] *Dirac Equation a*, folio 12 (page 11).

4.5 The Impossible "Picture"

So far so good, would it not be for the fact that this whole enterprise of devising "pictures" of quantum mechanical phenomena clashes with the education in quantum mechanics that Feynman had received. Like many other physicists of the period, Feynman had learned much about quantum mechanics from the textbook written by Dirac.[20] Feynman recalls using Dirac's book when he was an undergraduate, and describes the "challenge" posed by its very last sentence: "Some essentially new physical ideas are here needed."[21] Welton[22] also remembers that Feynman consulted Dirac's book while he was an undergraduate.

Dirac maintained that physicists had to be satisfied with the theory of quantum mechanics, even if it did not provide a full understanding of the mechanisms at work behind the phenomena described by the theory's equations:

> The methods of progress in theoretical physics have undergone a vast change during the present century. The classical tradition has been to consider the world to be an association of observable objects (particles, fluids, fields, etc.) moving about according to definite laws of force, so that one could form a mental picture in space and time of the whole scheme. This led to a physics whose aim was to make assumptions about the mechanism and forces connecting these observable objects, to account for their behaviour in the simplest possible way. It has become increasingly evident in recent times, however, that nature works on a different plan. Her fundamental laws do not govern the world as it appears in our mental picture in any very direct way, but instead they control a substratum of which we cannot form a mental picture without introducing irrelevancies. (Dirac 1935, p. vi)

Feynman most probably took another closer look at Dirac's textbook around the time that he wrote to Welton (that is, in 1947); after all, it was Dirac's equation with which Feynman was struggling. In addition, Feynman is very likely to have consulted Dirac's textbook when he wrote a critique on a paper presented by Dirac at the Princeton Bicentennial Conference held in September 1946.[23] Feynman was not at all impressed with the paper, because, according to him, it did not address current issues in electrodynamics. In particular, he did not appreciate the emphasis Dirac placed on the Hamiltonian—the same Dirac who had inspired the young Feynman to dispense with a Hamiltonian in his PhD work and to employ instead the Lagrange function using the action principle.[24]

Perhaps because Dirac's supposedly shallow paper annoyed him, Feynman was only too happy to ignore all of Dirac's caveats. The following passage, again from his letter to Welton, is probably a direct reaction to Dirac's claim, in the preface quoted above, that it is impossible to make a "mental picture in space and time" of quantum phenomena:

---

20 Dirac (1935).

21 Dirac (1935, p. 297); quoted in Feynman (1966, p. 699).

22 Welton (2007, p. 46).

23 Dirac (1946).

24 Weiner (1966, p. 15), see also Schweber (1994, p. 405).

> I dislike all this talk of there not being a picture possible but we only need know how to go about calculating any phenomena. True we only *need* calculate. But a picture is certainly a *convenience* & one is not doing anything wrong in making one up. It may prove to be entirely haywire while the equations are nearly right—yet for a while it helps. The power of mathematics is terrifying—and too many physicists finding they have correct equations without understanding them have been so terrified they give up trying to understand them. I want to go back & try to understand them. What do I mean by understanding? Nothing deep or accurate—just to be able to see some of the qualitative consequences of the equations by some method other than solving them in detail.[25]

The "terrifying" power of mathematics does not intimidate Feynman, though. He is determined to find "pictures" that help us "understand" the phenomena, rather than just correctly describe them using mathematical equations. However powerful a tool mathematics might be for deriving unanticipated mathematical expressions from other mathematical expressions, it will not yield physical claims unless the mathematical expressions are given a physical interpretation. Feynman's emphasis on the need to find a physical interpretation is also probably a response to Dirac's preface, in which Dirac states:

> Mathematics is the tool specially suited for dealing with abstract concepts of any kind and there is *no limit to its power* in this field. (Dirac 1935a, p. vii)

Feynman would not deny the usefulness of mathematics. However, this kind of power is simply not appropriate when it comes to trying to "straighten out" the conceptual problems of a physical theory.

The excerpt from Feynman's letter quoted above is perhaps slightly ambiguous on the subject of the importance of pictures and mathematical formulae. By using terms such as "convenience" and clauses such as "yet for a while it helps", Feynman is hedging his claims about the necessity of pictures and does not attach too much importance to them. His agenda, however, clearly shows the importance he attributes to the quest for a physical interpretation of the equations. In Welton's letter to Feynman, Welton describes the progress he is making in devising a theory of electrons and protons by means of the so-called second quantization (see Section 4.2). Feynman kindly feigns interest, but he does not seem genuinely keen to study a method that, by mathematical manipulations, gives correct results yet does not provide any physical interpretation of what the mathematical equations are supposed to describe.

> I will work on your 2nd quantization, when I get time. I really believe it is correct—but is another example of the terrifying power of math. to make us say things which we don't understand but are true.[26]

---

[25] *Dirac Equation a*, folio 12 (page 11), emph. in the original.

[26] *Dirac Equation a*, folio 16 (page 13, last page) see Fig. 4.18.

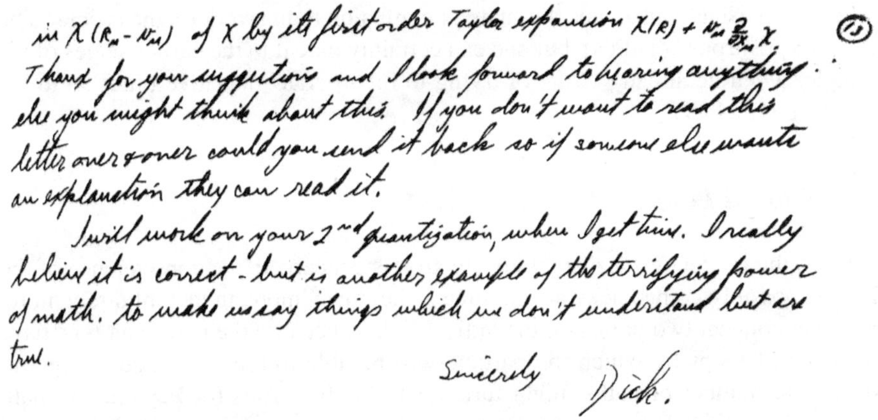
in $X(R_n - N_n)$ of X by its first order Taylor expansion $X(R) + N_n \frac{\partial}{\partial R_n} X$. (13)
Thank for your suggestions and I look forward to hearing anything else you might think about this. If you don't want to read this letter over & over could you send it back so if someone else wants an explanation they can read it.
I will work on your 2nd quantization, when I get time. I really believe it is correct - but is another example of the terrifying power of math. to make us say things which we don't understand but are true.
Sincerely Dick.

Fig. 4.18 Thirteenth and last page of Feynman's letter to Welton, February 1947 (*Dirac Equation a*, folio 16, reprinted with permission of Melanie Jackson Agency, LLC)

4.6 Positrons: A Threat to Path Counting

In Section 4.3.2 we saw that the generalization to three spatial dimensions of the description of the quivering electron did not satisfy Feynman. The treatment remained too "mathematical" and did not help him get any closer to his quest to "understand" the Dirac equation. As far as I can tell from the available source material, Feynman abandons his attempts at generalizing to three spatial dimensions after his correspondence with Welton ends.

Yet Feynman is preoccupied enough with the one-dimensional case. Up to now he has completely neglected one of the Dirac equation's most conspicuous features: it allows for solutions that apparently describe states with negative energy. By the time Feynman had begun to tackle the Dirac equation, the interpretation of these states as "holes" in the "sea" of electrons was quite firmly established, these "holes" describing positrons.[27] Feynman recalls that this established interpretation always troubled him and that he preferred describing positrons as electrons going backwards in time rather than as "holes":

> But one step of importance that was physically new was involved with the negative energy sea of Dirac, which caused me so much logical difficulty. I got so confused that I remembered Wheeler's old idea about the positron being, maybe, the electron going backward in time.[28]

---

[27] See, e. g., Dirac (1935, p. 270ff).

[28] Feynman (1966, p. 706) (middle column), see also p. 702 (middle and right column), p. 705 (right column); see also Schweber (1994, pp. 387/388) and Weiner (1966a, p. 31/32).

Feynman usually attributes this state of confusion to his work on the article "The theory of positrons" (ThPos), but one can certainly date it to the earlier stages of his struggle, when he already wrote of trying to incorporate "negative times" into his work (see Fig. 4.19).

4.6.1 Closed Loops

As far as the model of the zigzagging electron is concerned, the consequence of the electron possibly going backwards in time is that many more, in fact infinitely more, paths can connect two points on the lattice. This is because the paths can now have the shape of loops by which the particle will be able to return to a certain point. An infinite number of paths might turn out to be disastrous for Feynman's "path counting" solution. But it is not. Feynman can show that "any completely closed loop" will not contribute to the wave function, since every loop is part of two paths, one that goes "around [to the] left" and one that goes "around [to the] right" and that these two contributions cancel each other out.

Let us see how Feynman proceeds (see Fig. 4.19). He starts with the observation that, to solve the Dirac equation, corner turns, after which the path goes backwards in time, have to contribute a factor $-i\varepsilon$ instead of the $i\varepsilon$ for normal paths (number 1):

"Each turn to $+$ time gives $i\varepsilon$
$-$ $-i\varepsilon$."

Feynman draws the simplest of all loops, that is, one that goes "around [to the] left" (number 2, reproduced in Fig. 4.20), and determines that it has first to make a turn to $+$ time, then two turns to $-$ time and again a turn to $+$ time, yielding a contribution "$+i(-i)(-i)(+i)$". The path going "around [to the] right" contributes "$+i(+i)(-i)(+i)$". He writes one contribution over the other in order to compare them more easily. The two contributions differ only by the sign of the second factor. Feynman draws a line below the two contributions, as one usually does with an addition of numbers. The sum of the two contributions is zero, because the only difference is a sign. Feynman can, therefore, conclude below the horizontal line: "cancels".

The next drawing (see Fig. 4.21) makes the relationships more explicit. The loop that goes "around [to the] right" is in fact a step that yields, first, two positive factors, second, a negative factor and, third, a positive one. The absolute values of the right and left loops agree, but the contributions have the opposite sign and, therefore, cancel each other out.

4.6.2 Diagrammatic Induction

Any loop that Feynman can draw is only an instance of a closed loop for which the cancellation he claims is correct. So how can he prove the general statement that

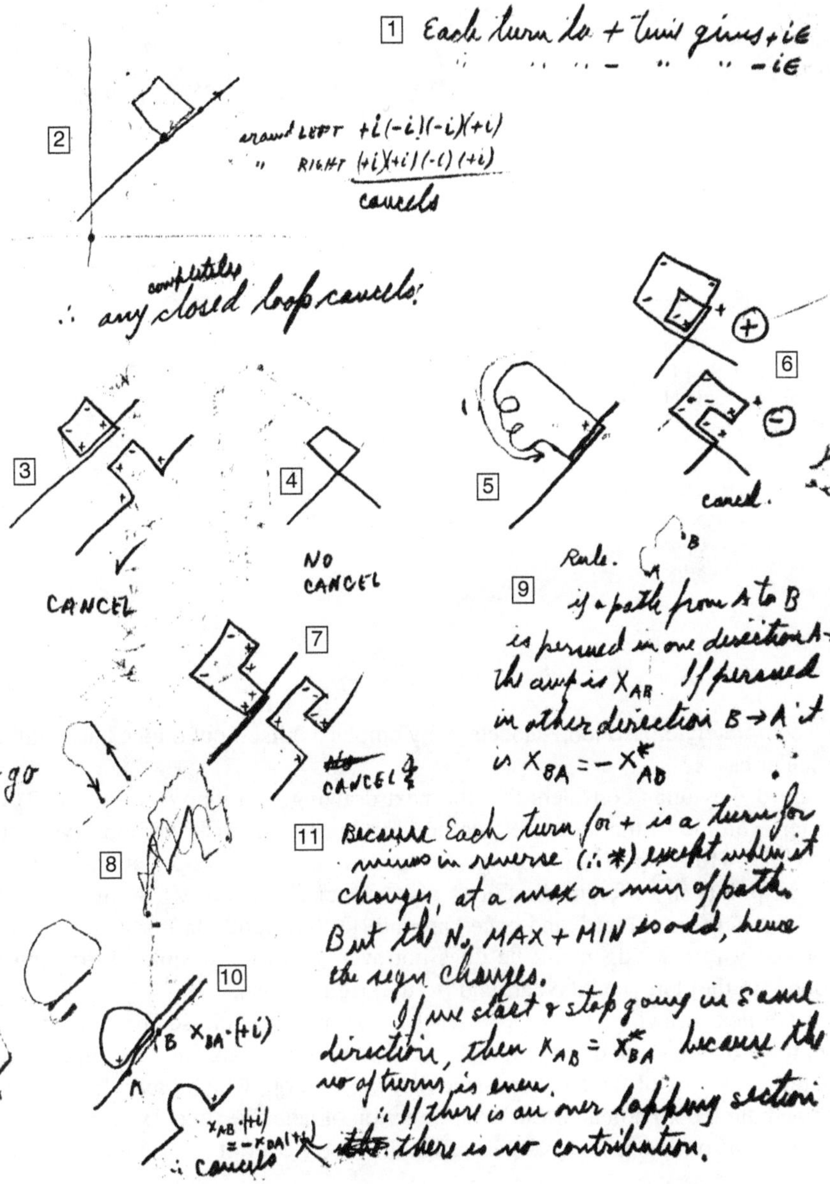

Fig. 4.19 The incorporation of "negative times" into the model of the quivering electron and proof that "any completely closed loop cancels" itself out, ca. 1947 (*Space-time approach to quantum electrodynamics*, folio 5, reprinted with permission of Melanie Jackson Agency, LLC)

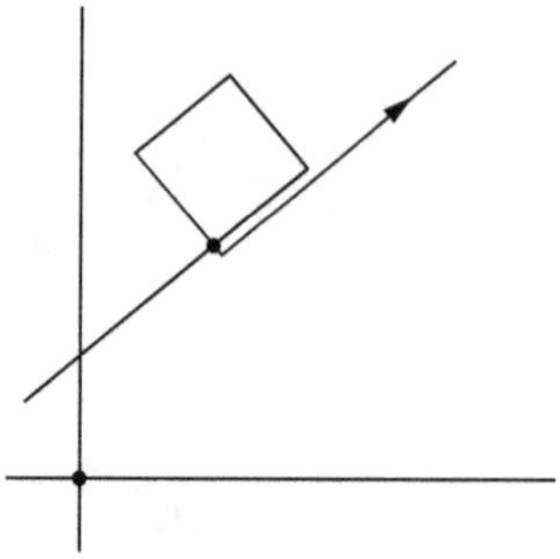

Fig. 4.20 Detail of Fig. 4.19 (number 2), adapted by the author (A. W.)

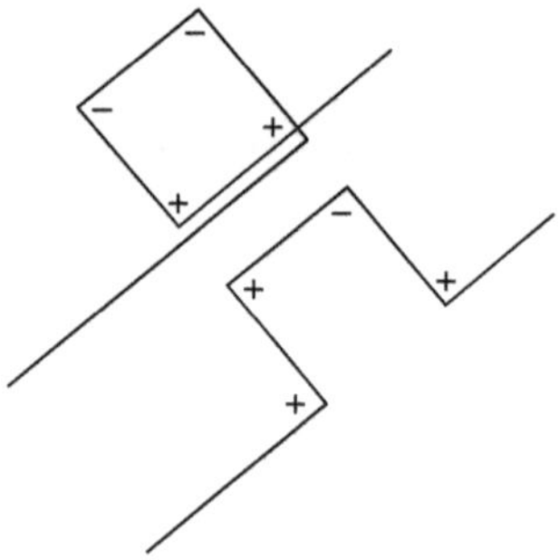

Fig. 4.21 Detail of Fig. 4.19 (number 3), adapted by the author (A. W.)

"*any* completely closed loop cancels" (my emphasis) itself out on the basis of these particular cases?

Indeed, Feynman considers that the next drawing of a loop (see Fig. 4.22) is a counter-example to the tentative rule he is trying to establish, and observes: "NO CANCEL". However, this is wrong! As far as I can see, a loop that has both start and end points in different directions also cancels itself out when circumscribed once "around [to the] left" and once "around [to the] right". In my view, Feynman is too hasty in his judgement; he does not even go to the trouble of drawing the "around [to the] left" and the "around [to the] right" loops.

In the next drawing (see Fig. 4.23) the first and last line segment of the loop point towards the same direction, and therefore they overlap. Here Feynman examines the variations of the left-hand-side part of the loop, which (apart from the two intermediate loops) points towards the direction of negative time. He considers how the variations of this backwards part affect the loop's overall contribution. Feynman

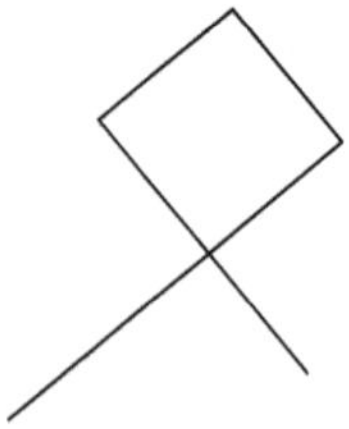

Fig. 4.22 Detail of Fig. 4.19 (number 4), adapted by the author (A. W.)

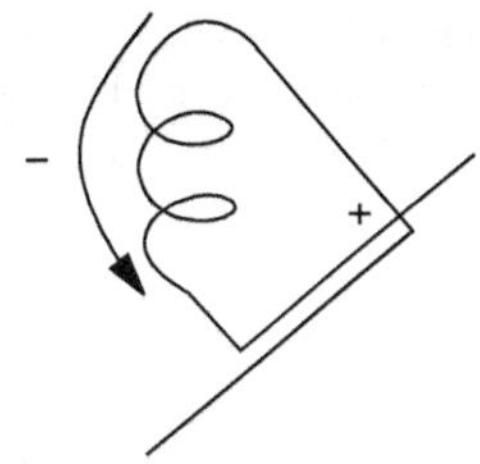

Fig. 4.23 Detail of Fig. 4.19 (number 5), adapted by the author (A. W.)

draws the diagram in order to convince himself that the inset loops can be ignored, since they contribute two minus signs (and two plus signs) and thus do not alter the sign of the rest of the loop. In order to determine the sign of the loop, it is, therefore, sufficient to consider the single backwards part (the curved arrow-headed line), which gives a minus sign (annotation left of the curved arrow-headed line).

The proof of the general statement is a proof akin to mathematical induction. The statement can be verified in some simple instances. Then a construction procedure is defined by which all the instances can be constructed from another instance. It then only remains to be shown that the construction procedure always leads from an instance for which the statement is true to an instance for which it is also true.

A closed loop, such as the one in Fig. 4.23, has an odd number of left corners and an even number of right corners or vice versa in its wavy backwards part, irrespective of how many loops the backwards part contains (see Fig. 4.24). When continuing "around [to the] left", the left corners contribute a negative sign and the right corners a positive sign. Therefore, the whole wavy backwards part with two loops contributes a negative sign (to the left of the wavy backwards part) when circumscribed "around [to the] left". "Around [to the] right" is simply the opposite: the left corners contribute a positive sign and the right corners a negative sign. Thus there is an even/odd number of negative signs "around [to the] left", while "around [to the] right" there is an odd/even number of negative signs. This means that in one case the overall sign is negative and in the other case it is not. Therefore, any closed

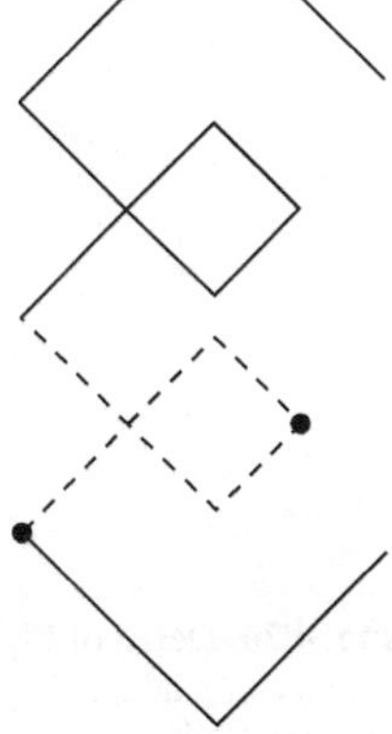

Fig. 4.24 The wavy backwards parts of closed loops contain an even number of left corners and an odd number of right corners. Every *inset* of a new loop adds one left corner and one right corner. Therefore, either the number of left corners is even and the number of right corners is odd or vice versa (Drawing by A. W.)

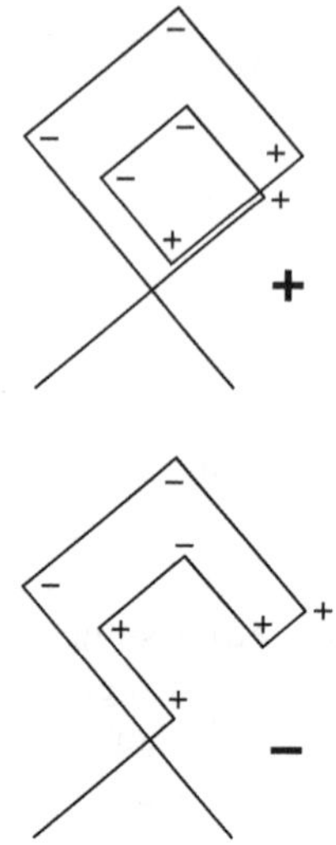

Fig. 4.25 Detail of Fig. 4.19 (number 6), adapted by the author (A. W.)

loop cancels itself out, since in the procedure of path counting any closed loop is circumscribed once "around [to the] left" and once "around [to the] right".

Feynman then returns to a more complex type of loop, that is, one whose initial and final directions differ (Fig. 4.25). After having (erroneously) inferred that the simplest form of such a loop does not cancel itself out (see above), he now (correctly) sees that this instance does indeed do so. However, he does not revise his conclusion of the simplest case.

Judging from the layout of his notes, Feynman next reconsiders an instance of the type with an overlapping segment (Fig. 4.26). Feynman still does not seem able to judge whether the instance confirms the proposed rule that "any completely closed loop cancels" itself out. Of this instance he first writes "NO CANCEL" but then corrects himself and (rightly) concludes that it does.

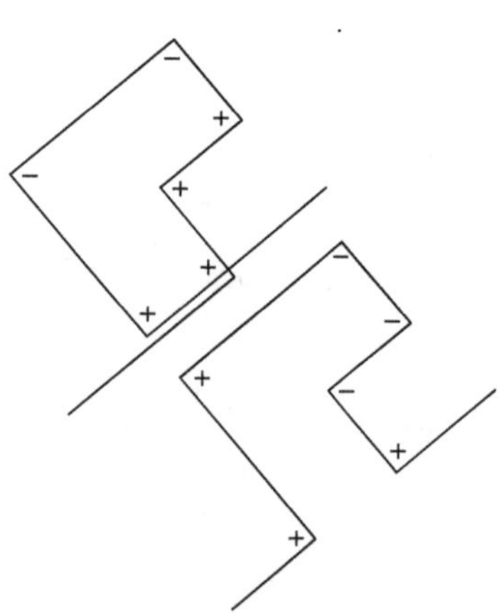

Fig. 4.26 Detail of Fig. 4.19 (number 7), adapted by the author (A. W.)

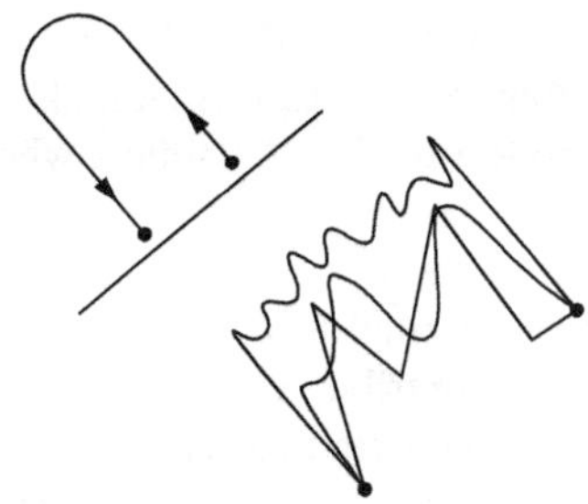

Fig. 4.27 Detail of Fig. 4.19 (number 8), adapted by the author (A. W.)

Then Feynman recognizes that the problem of determining the overall sign of the "around [to the] left" and "around [to the] right" loops can be reduced to the problem of determining the relative sign between the amplitude for describing a U-turn in one or the other direction (see diagram annotated "to go" and the accompanying scribbles representing the more general case of a U-turn with intermediate jagged lines [Fig. 4.27]). This is again a diagrammatic induction: the U-turn with no jagged lines is the simplest instance for which the rule is verified. The drawing of the jagged lines verifies that, in the inductive step, the truth of the statement is preserved when going from one instance to the next complex one.

Feynman is now ready for a more specific formulation of the general statement: the "Rule" (Fig. 4.19, number 9). With A and B denoting the two ends of the U-turn (see Fig. 4.28), the amplitude for going from B to A through the U-turn is minus the complex conjugate of the amplitude for going through it from A to B:

> if a path from A to B is persued [sic] in one direction $A \rightarrow B$, the amp[litude] is χ_{AB}. If persued [sic] in other direction $B \rightarrow A$ it is $\chi_{BA} = -\chi^*_{AB}$

Feynman then describes in words the diagrammatic proof of the rule. A left or right turn contributes a factor $+i$ to the amplitude χ when the turn travels towards positive time (upwards) and $-i$ when the turn travels towards negative time (downwards). That is, if the direction in which the particle is following the path is switched, these contributions will change sign or, what is the same for purely imaginary numbers such as i, are replaced with their complex conjugates. If the contribution from *each* turn were replaced with its complex conjugate, the reverse

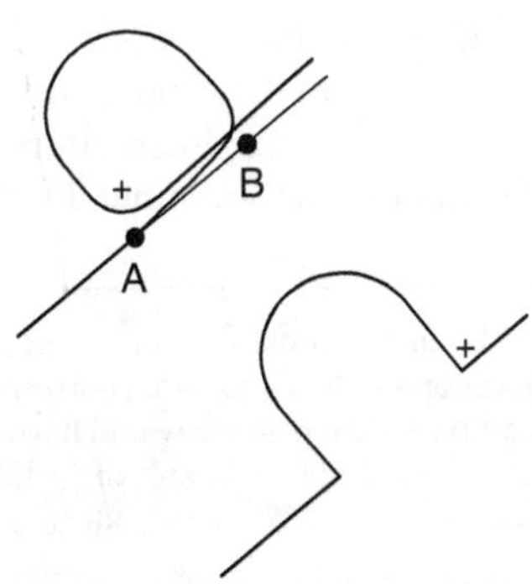

Fig. 4.28 Detail of Fig. 4.19 (number 10), adapted by the author (A. W.)

amplitude would be the complex conjugate of the original amplitude: $\chi_{BA} = \chi^*_{AB}$. However, the turns that make the path curve in the opposite time direction are turns "to + time" or "− time", whether they travel from A to B or from B to A. Therefore, the contributions of these extremal turns remain the same in χ_{BA} as in χ_{AB}. Since there is always an odd number of extremal turns in the loop (not counting the extremal turn, which brings an around to the left turn back onto the path) and their contribution is purely imaginary ($\pm i$), the fact that their contributions are not replaced with their complex conjugates, unlike those of the left/right turns, results in an additional overall minus sign (number 11), that is $\chi_{BA} = -\chi^*_{AB}$[29]; in Feynman's words:

> Because Each [sic] turn for + is a turn for minus in reverse ($\therefore *$) except when it changes, at a max or min of path. But the No. MAX + MIN is odd, hence the sign changes.

For the case of a loop inserted on a straight path, it then follows that the contributions of going "around [to the] left" and going "around [to the] right" cancel each other out, the reason being that this type of loop has an even number of turns. In particular the number of extremal turns is even, since we now count the turn that brings the around to the left loop back onto the path. There is, therefore, no additional minus sign and the reverse amplitude is just the complex conjugated original amplitude: $\chi_{AB} = \chi^*_{BA}$. Yet a product of an even number of purely imaginary factors (as the is are) is real and since we now have to count the turns that bring the loops back onto the path (see the + sign in the drawings near number 10), we have an additional factor $+i$, which changes sign when it is replaced with its complex conjugate: hence $\chi_{AB} = -\chi_{BA}$. Therefore, these inserted loops can be ignored in path counting, since the two contributions they would yield cancel each other out:

> If we start & stop going in same direction, then $\chi_{AB} = \chi^*_{BA}$ because the no[.] of turns is even. $\therefore$ If there is an over lapping section there is no contribution.

4.7 Direct Interaction

After having discovered how to circumvent the self-inflicted difficulty of paths going backwards in time, Feynman still needs to work out the essential features of the model of the quivering electron. To "straighten out some of the present troubles" of QED, he has to be able to describe not only a single electron but at least two electrons and their interactions.

However, the interaction description of the Dirac equation does not fit into Feynman's programme. In the Dirac equation, as in classical electrodynamics,

---

[29] In other words: let $\chi_{AB} = ab$, where a contains the factors that correspond to the extremal turns and b the factors that correspond to the left/right turns. If each factor were replaced with its complex conjugate we would have $\chi_{BA} = \chi^*_{BA} = a^*b^*$. Now, because the factors that correspond to the extremal turns are, in fact, not replaced with their complex conjugates, we have instead $\chi_{BA} = ab^* = \frac{ab^*}{a^*b^*}a^*b^*$. Since a is a product of an odd number of is, we have $a = e^{i\frac{\pi}{2}(2n+1)}$, where n is an integer, and, therefore, $\chi_{BA} = e^{i\pi(2n+1)}a^*b^* = -\chi^*_{AB}$.

interaction is only taken into account indirectly: one of the charges generates an electromagnetic potential and that potential acts on the other particle. Feynman (and Wheeler) had to dispose of this feature in order to be able to eliminate the uninterpretable divergences of the theory. So, although Feynman has, up to now, been busy interpreting an equation that he had adopted from Dirac, he now has to find a yet unknown equation—an equation, that is, that would describe the *direct* interaction of two electrons, akin to the classical theory of Wheeler and Feynman.[30]

In order to continue tackling Dirac's equation, which consists of interpreting the differential equation for the quantum mechanical wave function that describes the motion of a quivering electron, Feynman must obtain differential equations from the classical action of his and Wheeler's theory. This amounts to constructing a Hamiltonian, which then yields the differential equations via $\frac{\hbar}{i}\frac{\partial\psi}{\partial t}$. (With a non-relativistic Hamiltonian this gives Schrödinger's equation; with the Dirac's Hamiltonian Dirac's equation.) In his thesis, Feynman tried to circumvent such a procedure, since the Wheeler–Feynman theory had an action but not a classical Hamiltonian.

In his thesis, Feynman had solved the problem of how to quantize a theory whose "equations of motion of the particles are expressed classically as a consequence of a principle of least action, and cannot, it appears, be expressed in Hamiltonian form".[31] The alternative method of quantization that Feynman developed in his thesis can also be applied to systems of more than one particle. For these cases his method has the particular advantage over canonical quantization in that "the form of Schrödinger's equation which will be arrived at will be definite and will not suffer from the type of ambiguity one finds if one tries to substitute $\frac{\hbar}{i}\frac{\partial}{\partial q}$ for p_q in the classical Hamiltonian".[32]

Feynman recalls having been aware of the fact that, in principle, it might have been possible to express the theory by a Hamiltonian and to quantize it canonically. But when he was writing his thesis, the advantages of not invoking a Hamiltonian had made him not even attempt it:

> The standard method of going to quantum mechanics from classical mechanics assumed there was a Hamiltonian. Well, there wasn't—in this form. If I expressed it in the terms of fields or something, there might have been. I was very reluctant to do that, maybe incorrectly. There probably was another way around. But I insisted always to represent only the particles. (Weiner 1966b, p. 146, reprinted with permission. Copyright 1966, American Institute of Physics)

Now that he needs to find differential equations for the quivering electron, Feynman reconsiders the possibility of constructing a Hamiltonian that would describe direct particle interaction and then quantizes it using the canonical procedure.

---

30 "Minutes of the Cambridge, Massachusetts, Meeting, February 21 and 22, 1941" 1941; Wheeler and Feynman (1945, 1949).

31 Feynman (2005, p. 5).

32 Feynman (2005, p. 31).

Although Feynman is still using an action function that does not introduce any fields but only describes the particles, he does now try to transform the action function into a classical Hamiltonian. The problem here is that, in the Wheeler–Feynman theory, the action depends on different times. As we will see in the next section, Feynman tries to circumvent this problem by introducing new variables, similar to the "light-cone coordinates" introduced at the beginning of his struggle with the Dirac equation (see Section 4.1.2).

4.7.1 An Attempt to Construct a Hamiltonian

Feynman adopted the action of the Wheeler–Feynman theory from Karl Schwarzschild (1903), H. Tetrode (1922) and A. D. Fokker (1929).[33] In order to compare the known situation of classical electrodynamics with the situation with which he is struggling, Feynman writes down how he would treat the interaction of two particles in one (spatial) dimension in the classical framework (see Fig. 4.29, number 1):

$$\begin{aligned} S = &\int \sqrt{\dot{t}^2 - \dot{x}^2}\, d\alpha + \int \sqrt{\dot{\tau}^2 - \dot{\xi}^2}\, d\beta \\ &+ k \int\int \delta((t-\tau)^2 - (x-\xi)^2)(\dot{t}\dot{\tau} - \dot{x}\dot{\xi})\, d\alpha\, d\beta. \end{aligned} \tag{4.20}$$

The coupling constant, which determines the strength of the interaction, is k , which depends on the charges of the two particles: it is $k = e_a e_b/c$, where e_a denotes the charge of particle a, and e_b that of b; c is the speed of light. (x, t) are the space-time coordinates of the first particle, (ξ, τ) those of the second particle. α and β are the respective proper times of the particles.[34]

As already mentioned, the differential equation for the wave function is determined by the Hamiltonian system. Usually, the action is simply the time integral of the Lagrange function of the system, and from the Lagrangian one can obtain the Hamiltonian using a Legendre transformation. As Feynman emphasized in his thesis (Feynman 2005a) and in RMP48 (and was also explicitly mentioned in, for example, Tetrode 1922, p. 320), the problem here is that the action cannot be written as a simple time integral and, therefore, "cannot be described directly in Hamiltonian form".[35]

As with the Dirac equation for a single particle (see Section 4.1.2), Feynman introduces coordinates that are rotated by 45° with respect to the ordinary coordinate system (x-t for the first particle, ξ-τ for the second particle; see Fig. 4.29, number 2):

---

[33] See Wheeler and Feynman (1949, p. 425).

[34] See Wheeler and Feynman (1949), equation (1).

[35] Feynman (2005, p. 10).

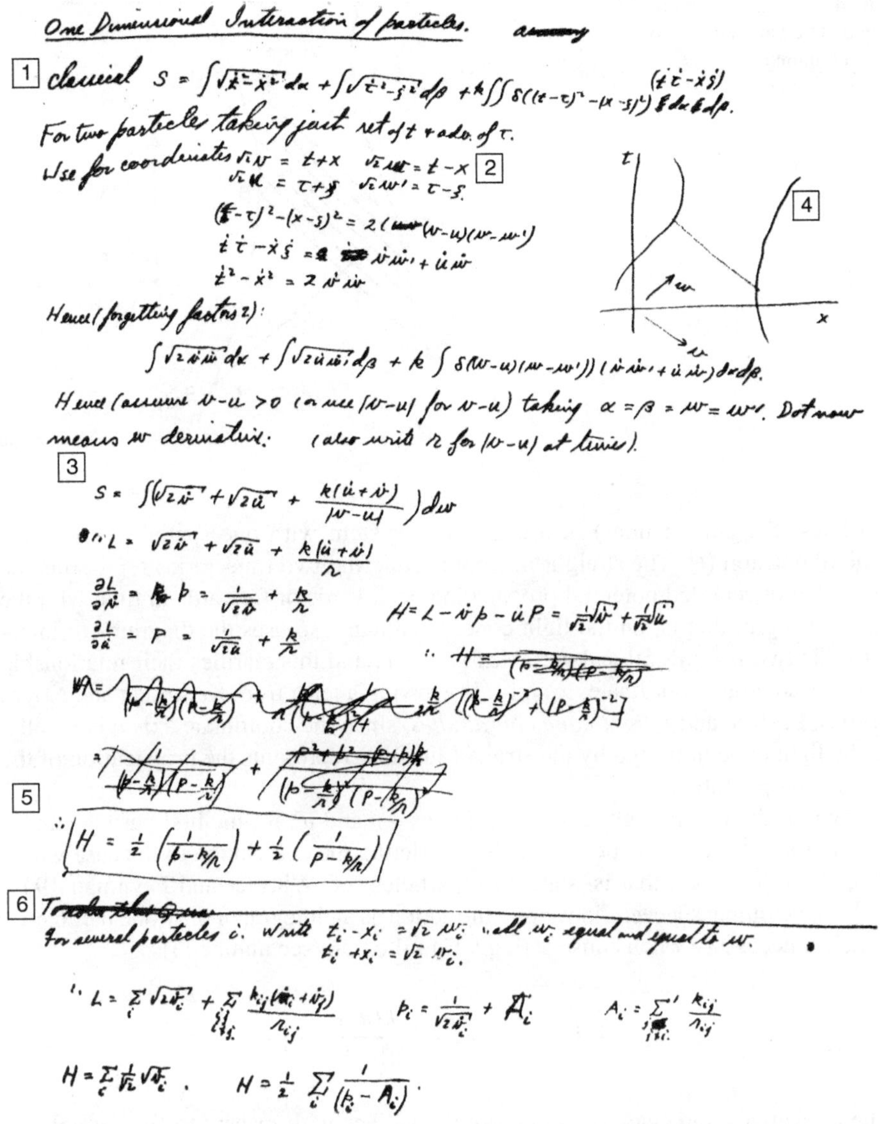

One Dimensional Interaction of particles.

1 classical $S = \int \sqrt{\dot{t}^2 - \dot{x}^2}\, d\alpha + \int \sqrt{\dot{\tau}^2 - \dot{\xi}^2}\, d\beta + k \iint \delta((t-\tau)^2 - (x-\xi)^2)\, (\dot{t}\dot{\tau} - \dot{x}\dot{\xi})\, d\alpha\, d\beta.$

For two particles taking just ret of t + adv. of τ.

Use for coordinates $\sqrt{2}v = t + x$ $\sqrt{2}w = t - x$ 2

$\sqrt{2}u = \tau + \xi$ $\sqrt{2}w' = \tau - \xi$.

$(t-\tau)^2 - (x-\xi)^2 = 2(v-u)(w-w')$

$\dot{t}\dot{\tau} - \dot{x}\dot{\xi} = \dot{v}\dot{w}' + \dot{u}\dot{w}$

$\dot{t}^2 - \dot{x}^2 = 2\dot{v}\dot{w}$

Hence (forgetting factors 2):

$\int \sqrt{2\dot{v}\dot{w}}\, d\alpha + \int \sqrt{2\dot{u}\dot{w}'}\, d\beta + k \int \delta((v-u)(w-w'))\, (\dot{v}\dot{w}' + \dot{u}\dot{w})\, d\alpha\, d\beta.$

Hence (assume $v - u > 0$ or use $|v-u|$ for $v-u$) taking $\alpha = \beta = w = w'$. Dot now means w derivative. (also write r for $|v-u|$ at times).

3

$S = \int \left(\sqrt{2\dot{v}} + \sqrt{2\dot{u}} + \frac{k(\dot{u}+\dot{v})}{|v-u|} \right) dw$

$\therefore L = \sqrt{2\dot{v}} + \sqrt{2\dot{u}} + \frac{k(\dot{u}+\dot{v})}{r}$

$\frac{\partial L}{\partial \dot{v}} = p = \frac{1}{\sqrt{2\dot{v}}} + \frac{k}{r}$ $H = L - \dot{v}p - \dot{u}P = \frac{1}{\sqrt{2}}\sqrt{\dot{v}} + \frac{1}{\sqrt{2}}\sqrt{\dot{u}}$

$\frac{\partial L}{\partial \dot{u}} = P = \frac{1}{\sqrt{2\dot{u}}} + \frac{k}{r}$

5

$\therefore H = \frac{1}{2}\left(\frac{1}{p - k/r}\right) + \frac{1}{2}\left(\frac{1}{P - k/r}\right)$

6

For several particles i. Write $t_i - x_i = \sqrt{2}\, w_i$ all w_i equal and equal to w.

$t_i + x_i = \sqrt{2}\, v_i$.

$\therefore L = \sum_i \sqrt{2\dot{v}_i} + \sum_{i,j} \frac{k_{ij}(\dot{v}_i + \dot{v}_j)}{r_{ij}}$ $p_i = \frac{1}{\sqrt{2\dot{v}_i}} + A_i$ $A_i = \sum_{j}' \frac{k_{ij}}{r_{ij}}$

$H = \sum_i \frac{1}{\sqrt{2}}\sqrt{\dot{v}_i}$. $H = \frac{1}{2}\sum_i \frac{1}{(p_i - A_i)}$.

4

Fig. 4.29 The classical direct interaction of two particles in one spatial dimension, ca. 1947 (*Dirac Equation n*, folio 5, reprinted with permission of Melanie Jackson Agency, LLC)

$$\sqrt{2}v = t + x, \qquad \sqrt{2}w = t - x, \tag{4.21}$$

$$\sqrt{2}u = \tau + \xi, \qquad \sqrt{2}w' = \tau - \xi. \tag{4.22}$$

The diagram (Fig. 4.29, number 4) recalls the classical framework from which the action function was adopted. Similar diagrams are used, for example in Wheeler and Feynman (1949, p. 431); see also Fig. 4.30. It displays the trajectories of two

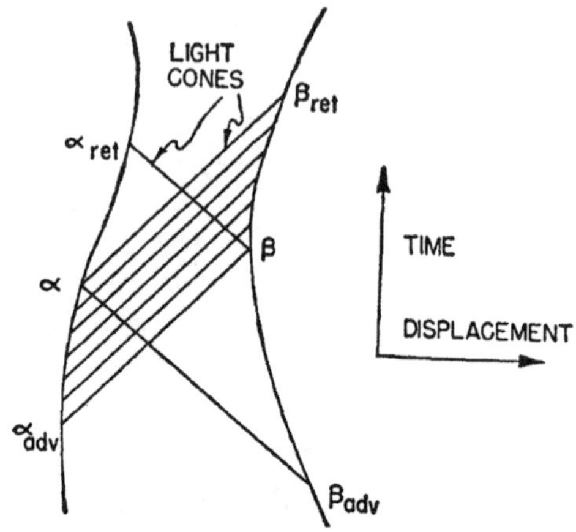

Fig. 4.30 Representation of the retarded and advanced potential mediating the interaction between two electrons, according to the Wheeler–Feynman theory of electrodynamics (Wheeler and Feynman 1949, p. 431)

particles (the curved lines) in a coordinate system with one spatial (x) and one time dimension (t). The straight line connecting the two trajectories represents the advanced or retarded potential propagating backwards or forwards in time with the speed of light, that is, on the light cone. Feynman also uses the diagrams to introduce the two unconventional variables w and v, and this clarifies their relationship with the standard coordinates x and t. Borrowing a term from Kauffman and Noyes (1996), I call w and v *light-cone coordinates,* since the coordinate axis v is parallel to the light cone indicated by the straight line that represents the propagation of the interaction potential.

By introducing the light-cone coordinates—v and w, for the first particle, and u and w' for the second particle—and by restricting the description to the case where $\alpha = \beta = w = w'$, that is, stationary particles (cf. Wheeler and Feynman 1949, p. 430), Feynman succeeds in writing the action as an integral of only one parameter, which is necessary when constructing a Hamiltonian (see number 3):

$$S = \int \left(\sqrt{2\dot{v}} + \sqrt{2\dot{u}} + \frac{k(\dot{u} + \dot{v})}{|v - u|} \right) dw. \tag{4.23}$$

The derivatives, indicated by a dot, are derivatives with respect to the variable w: "Dot now means w derivative" (just above number 3), that is, $\dot{u} = \frac{\partial u}{\partial w}$, $\dot{v} = \frac{\partial v}{\partial w}$.

Feynman can now read off the Lagrangian L from the action S, taking S as the integral of L over w, instead of the ordinary one time parameter t (the line just below number 3):

$$L = \sqrt{2\dot{v}} + \sqrt{2\dot{u}} + \frac{k(\dot{u} + \dot{v})}{r}, \tag{4.24}$$

where $r = |v - u|$.

Then Feynman can form the canonical momenta and construct the Hamiltonian from them and the Lagrangian (the next two lines):

$$\frac{\partial L}{\partial \dot{v}} = p = \frac{1}{\sqrt{2\dot{v}}} + \frac{k}{r}, \tag{4.25}$$

$$\frac{\partial L}{\partial \dot{u}} = P = \frac{1}{\sqrt{2\dot{u}}} + \frac{k}{r}, \tag{4.26}$$

$$H = L - \dot{v}p - \dot{u}P = \frac{1}{\sqrt{2}}\sqrt{\dot{v}} + \frac{1}{\sqrt{2}}\sqrt{\dot{u}}. \tag{4.27}$$

Finally, the Hamiltonian has to be expressed in terms of the canonical momenta p and P, which is Feynman's final and important (therefore boxed) result (number 5):

$$\boxed{H = \frac{1}{2}\left(\frac{1}{p - k/r}\right) + \frac{1}{2}\left(\frac{1}{P - k/r}\right)} \tag{4.28}$$

The symbol "∴" indicates that the Hamiltonian is a derived result. Beneath the horizontal line that he has drawn across the lower half of the page, Feynman generalizes his findings to the case of "several particles" (number 6).

With this Hamiltonian in hand Feynman can now proceed to the canonical quantization procedure by promoting the Hamilton function to the Hamilton operator that characterizes a quantum system (see Fig. 4.31). Feynman follows the usual procedure by which the canonical momenta (p and P) are replaced with a (spatial) derivative, in terms, however, of his new variables u and v. These are the light-cone

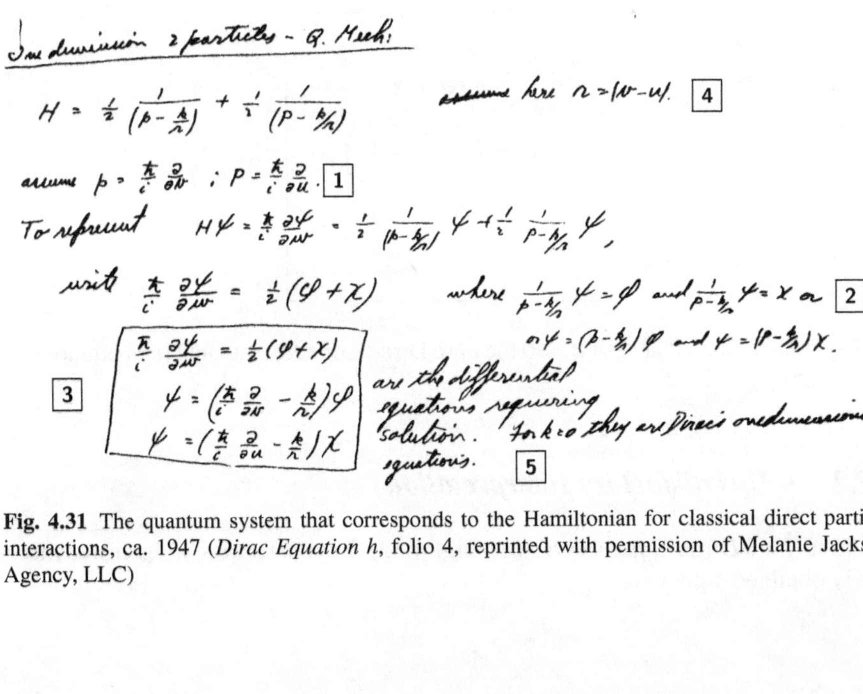

Fig. 4.31 The quantum system that corresponds to the Hamiltonian for classical direct particle interactions, ca. 1947 (*Dirac Equation h*, folio 4, reprinted with permission of Melanie Jackson Agency, LLC)

coordinates introduced earlier (see 4.21, 4.22), which mix space and time coordinates. Thus Feynman's canonical momenta read as follows (see number 1):

$$p = \frac{\hbar}{i}\frac{\partial}{\partial v}, \qquad P = \frac{\hbar}{i}\frac{\partial}{\partial u}, \tag{4.29}$$

which would give, as the quantum Hamilton operator:

$$H = \frac{1}{2}\left(\frac{1}{\frac{\hbar}{i}\frac{\partial}{\partial v} - k/r}\right) + \frac{1}{2}\left(\frac{1}{\frac{\hbar}{i}\frac{\partial}{\partial u} - k/r}\right). \tag{4.30}$$

However, in this expression derivatives appear in the denominator, which makes the definition of H as an operator unclear. One might try to define an operator by developing the expression in a power series, such that the derivatives would appear in an ordinary position. Feynman does not even write this Hamiltonian down but tries to circumvent the difficulty by defining two new wave functions ϕ and χ (number 2):

$$\frac{1}{(p - k/r)}\psi = \phi, \tag{4.31}$$

$$\frac{1}{(P - k/r)}\psi = \chi, \tag{4.32}$$

where ψ is the wave function of the whole two-particle system and $r = |v - u|$ (see number 4). The resulting differential equations for ψ, χ and ϕ are (number 3):

$$\boxed{\begin{aligned} \frac{\hbar}{i}\frac{\partial\psi}{\partial w} &= \frac{1}{2}(\phi + \chi), \\ \psi &= \left(\frac{\hbar}{i}\frac{\partial}{\partial v} - \frac{k}{r}\right)\phi, \\ \psi &= \left(\frac{\hbar}{i}\frac{\partial}{\partial u} - \frac{k}{r}\right)\chi. \end{aligned}}$$

Feynman claims that "For $k = 0$ they are Dirac's onedimensional [sic] equations" (number 5).

4.7.2 An Unsatisfactory Interpretation

On folio 2 of *Dirac Equation h*, (see Fig. 4.32), Feynman tries to interpret these newly obtained equations.

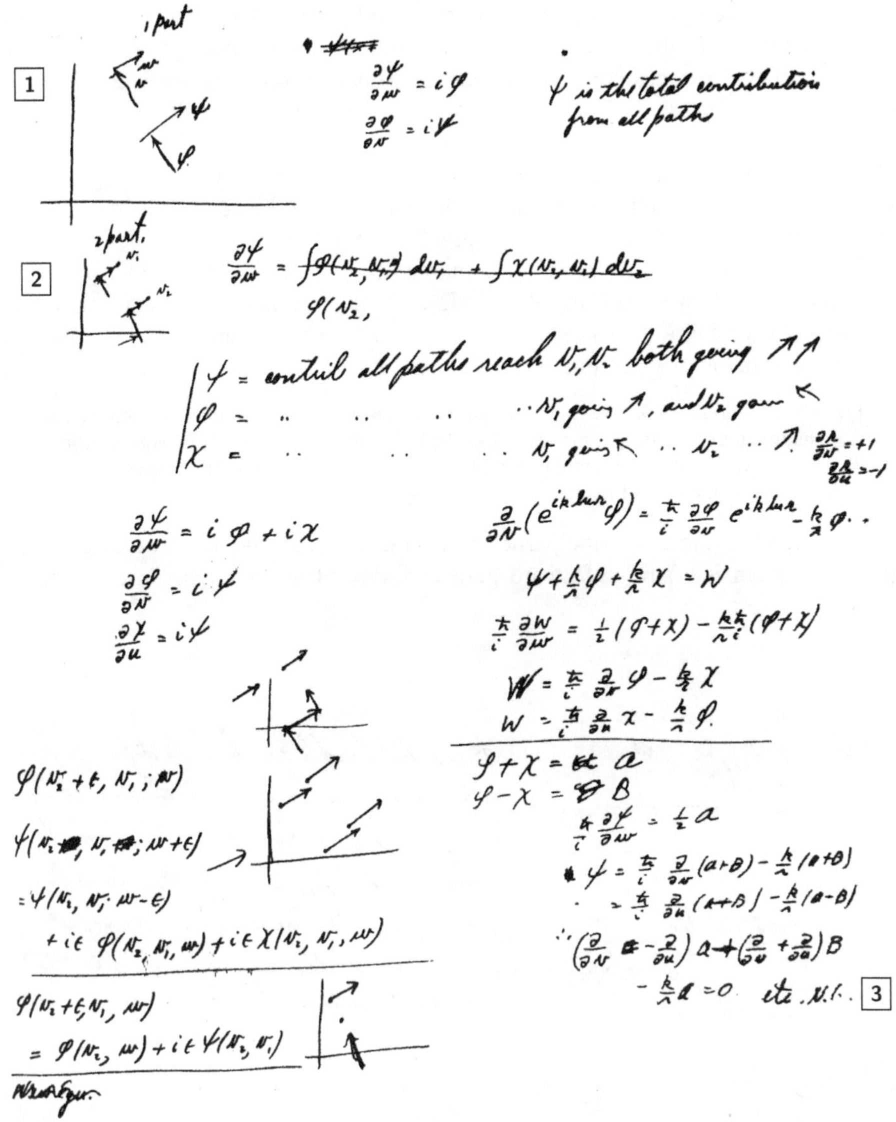

Fig. 4.32 Feynman's interpretation of the "Dirac" equation for two non-interacting particles, ca. 1947 (*Dirac Equation h*, folio 2, reprinted with permission of Melanie Jackson Agency, LLC)

If χ is set to zero (and $k = 0$, no interaction), the resulting equations have the same form as Feynman's version of the Dirac equation for one particle, where $w = u$, $\psi = \psi_R$ and $\phi = \psi_L$. The diagram entitled "1 part[icle]" (see number 1) again shows an interpretation of the right- and left-moving components of the wave function. The second diagram (number 2), entitled "2 part.", shows the interpretation for two particles, that is, $\chi \neq 0$. In this case, ψ is the contribution from all pairs of paths by which particle 1 reaches point v_1 and particle 2 reaches point v_2, both particles moving to the right. Then ϕ is the contribution from all pairs of paths by which particle 1 reaches point v_1 and particle 2 reaches point v_2, particle 1 moving to the right and particle 2 moving to the left. The same goes for χ, except that particle 1 moves to the left and particle 2 to the right.

At the bottom of the right-hand-side column of Fig. 4.32, Feynman again tries to treat the case *with* interaction ($k \neq 0$). This, however, leads him to make false statements: "[...] = 0. etc. N.G.[=no good?!]" (number 3).

Feynman's attempts to incorporate interaction into his model of the quivering electron causes him serious problems. He finds that he can only extend his interpretation of the Dirac equation to interacting particles in a special case, which is entirely unsatisfactory (see Fig. 4.33, number 1):

> It is a bit hard to see how to define Φ for path pair AB and CD, since there are some terms from interaction at x from y which is unspecified. However if the interaction is zero beyond P we are OK. Hence, at present, I can only specify Φ for paths which are long enough that they go beyond the time of interaction (this stinks).

After having interacted, the particles are free and the wave function of the two-particle system is a straightforward generalization of the one-particle case (number 2):

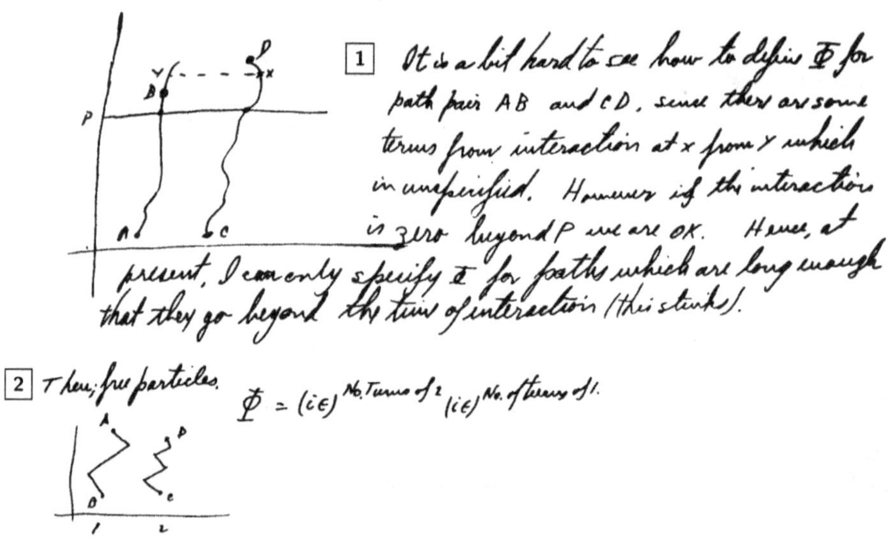

Fig. 4.33 Feynman's attempts to define the contributions of paths to the wave function for two interacting particles causes him serious problems. He explains that he can only treat a special case in this way, and this, he writes, "stinks", ca. 1947 (*Dirac Equation h*, folio 3, reprinted with permission of Melanie Jackson Agency, LLC)

$$\Phi = (i\varepsilon)^{\text{No.Turns of 2}}(i\varepsilon)^{\text{No.Turns of 1}} \tag{4.33}$$

However, this result is inadequate if one's aim is to "straighten out" some of QED's "present troubles". Feynman will, therefore, shortly abandon the model of the quivering electron. Clearly, not being able to treat interactions and free particles in three dimensions makes it impossible for him to continue working on the model.

Chapter 5
Free Propagation and Successive Scattering

5.1 Basic Solutions Not Basic Equations

From the time of his thesis and during his struggle to understand the Dirac equation, Feynman is preoccupied with finding the quantity that will determine the evolution of the wave function. While working on his thesis, he finds that the time evolution of the wave function is determined by the classical action, which is integrated along the possible paths that connect the start and end points of the particle. There is no classical action in the Dirac equation's description of quantum electrodynamics (QED), since the equation introduced a new degree of freedom: spin. In RMP48, Feynman constructs an action that yields the Dirac equation using his quantization method. However, because he cannot justify the action independently, Feynman considers this treatment to be "purely formal" and unsatisfactory (see Section 3.5).

To find and justify an action for a system described by the Dirac equation, Feynman adopts the model of the quivering electron (see Section 4.1). He looks for a description of the paths of the electron's *Zitterbewegung*[1] such that the Dirac equation will hold. That is, Feynman quantitatively describes the paths of the quivering electron, so that the sum of the quantities of each path will yield the quantity that determines the evolution of the wave function. Feynman is successful when working in one dimension without the interaction of particles. The description of each change in direction of the electron by a factor $i\varepsilon$ yields the desired quantity, a Bessel function in this case, when summed over all possible paths. He is even able to treat positrons successfully when he conceives of them as electrons going backwards in time (see Section 4.6).

Yet electrons in ordinary three-dimensional space and the interaction of electrons in one dimension causes him so many problems that he abandons his search to justify the quantity that describes the evolution of the wave function (see Sections 4.3 and 4.7). So, while working on his "Theory of Positrons" (ThPos), Feynman begins to consider the quantity as fundamental and tries to formulate an alternative theory of quantum mechanics, non-relativistic and relativistic, that is based on that quantity.

---

[1] Schrödinger 1930.

A. Wüthrich, *The Genesis of Feynman Diagrams*, Archimedes 26,
DOI 10.1007/978-90-481-9228-1_5, © Springer Science+Business Media B.V. 2010

Thus Feynman gives up trying to find a *microscopic* explanation of the quantity that describes the evolution of the wave function. During the Second World War, when Feynman was busy solving diffusion equations,[2] he learned that an alternative description of differential equations does not have to be a microscopic derivation of them. He learned (or, if he knew it already from his studies, he learned the importance of) how to describe diffusion phenomena not by a differential equation but by the *Green's function* associated with the equation.

As in classical mechanics, a Green's function is a particular solution to the differential equation with which it is associated. It is the solution to the inhomogeneous equation, where the inhomogeneity is given by a δ function. Generally speaking, a δ function has the value zero everywhere except in the case of one of its arguments, where the value is infinite but in a way that the integral of the function over the whole range of its argument is 1. For example, the δ function is often used to describe a "point source" of electricity or of a water wave. For the Schrödinger equation, $i\partial\psi/\partial t = H\psi$, the Green's functions are defined as the solutions to

$$(i\partial/\partial t - H)\psi(x,t) = i\delta(x - x_0)\delta(t - t_0) \tag{5.1}$$

(cf. equations (4) and (12) in ThPos). Usually there are several Green's functions associated with an equation and they are distinguishable from each other by their boundary conditions. Important types of boundary conditions include whether the Green's function vanishes for values of the variables that are greater or smaller than a particular given value.

Using a Green's function as the basic ingredient of a theory means placing the *solutions* to the differential equations higher up in the axiomatic system than the actual equations:

> The main principle is to deal directly with the solutions to the Hamiltonian differential equations rather than with these equations themselves. (ThPos, p. 749)

To establish a differential equation one usually derives it from a microscopic model of the situation. For instance, the differential equations describing the waves in a piece of string can be deduced from the considerations made of the stresses and strains in an infinitesimal piece of the string. Similarly, Feynman tries, without success though, to interpret Dirac's equations using the model of the quivering electron. However, from the time that he begins working on ThPos, he uses a relationship between two or more Green's functions (see, for example, 5.15) to describe the physical system and to endow that relationship, and not the differential equation, with a physical interpretation. The physical interpretation thus comes in at a more abstract level of the description.

---

[2] See, e. g., Galison 1998.

5.2 Interpreting the Schrödinger and Dirac Equations

To introduce his alternative formulation through the solutions to equations rather than through the equations themselves, Feynman applies the method first to Schrödinger's equation (of non-relativistic quantum mechanics)[3]:

$$i\partial\psi/\partial t = H\psi. \tag{5.2}$$

The general solution to Schrödinger's equation can be written as an integral over an initial wave function weighted by the Green's function $K(2, 1)$ associated with the equation[4]:

$$\psi(x_2, t_2) = \int K(2, 1)\psi(x_1, t_1)\, d^3x_1. \tag{5.3}$$

Rather than try to find a microscopic derivation of K, Feynman expresses K in terms of the *eigenvalues* and *eigenfunctions* of the Hamiltonian H. The eigenfunctions of the Hamiltonian H are those wave functions ϕ_n for which

$$H\phi_n = E_n\phi_n \tag{5.4}$$

holds. The Es are the eigenvalues of H and, if $\int \psi^*(x, t)\psi(x, t)\, d^3x$ is finite, the possible values for the energy of the system. K can be expressed by the eigenvalues and eigenfunctions as explained in the following paragraph.

As it is a solution to the Schrödinger equation, $\psi(x_2, t_2)$ can be written as $\psi(x_2, t_2) = \exp(-iH(t_2 - t_1))\psi(x_2, t_1)$, where $\psi(x_2, t_1)$ is the wave function at some other time t_1. Since the eigenfunctions, ϕ_n, form a basis for the space of the wave functions, $\psi(x_2, t_1)$ can be expressed as a superposition of the (stationary, that is, time-independent) eigenfunctions: $\psi(x_2, t_1) = \sum_n C_n(t_1)\phi_n(x_2)$. This implies that $\psi(x_2, t_2) = \exp(-iH(t_2 - t_1)) \sum_n C_n(t_1)\phi_n(x_2)$. Since $\phi_n(x_2)$ are eigenfunctions of H, $H\phi_n$ can be replaced with $E_n\phi_n$ (see (5.4)). The coefficients $C_n(t_1)$ are given by the scalar product in the space of the wave functions: $C_n(t_1) = \int \phi^*(x_2)\psi(x_2, t_1)\, d^3x_2$. Here x_2 is a dummy variable and can, therefore, be replaced with any other variable, in particular with x_1. We then obtain:

$$\psi(x_2, t_2) = \sum_n \exp(-iE_n(t_2 - t_1)) \int \phi_n^*(x_1)\psi(x_1, t_1)\, d^3x_1\, \phi_n(x_2).$$

If we compare this equation with (5.3), we can read off $K(2, 1)$ expressed in terms of the energy eigenfunctions and eigenvalues of the physical system[5]:

---

[3] Feynman uses units where $\hbar = 1$.

[4] See ThPos, equation (2).

[5] ThPos, equation (3).

$$K(2,1) = \sum_n \phi_n^*(x_1)\phi_n(x_2)\exp(-iE_n(t_2 - t_1)). \tag{5.5}$$

This equation is only valid for $t_2 > t_1$. For $t_1 > t_2$, it is "convenient"[6] to define $K(2,1) = 0$.

If the Schrödinger equation describes the behaviour of a single electron, the wave function gives the probability amplitude for localizing the electron at a particular space-time point. $K(2,1)$ can then be "[called] the total amplitude for arrival [of the electron] at x_2, t_2 starting from x_1, t_1" (p. 750). However, K does *not* describe a space-time path of the electron from (x_1, t_1) to (x_2, t_2), but "results from adding an amplitude, $\exp iS$, for each space time path between these points, where S is the action along the path [RMP48, p. 750]".

The relativistic quantum mechanics of electrons and their interaction, the central phenomena of QED, is described by the Dirac equation. The by-then established method for quantizing electrodynamics was a procedure that involved a "second" quantization. That is, one took the quantum mechanical equation for the wave function ψ that is supposed to describe the physical system and, in a second step, declared ψ to be an *operator* that acts on a Hilbert space of states, rather than letting the wave function itself describe a state of the system.

Feynman's description of QED is different. He does not adopt the process of second quantization but keeps to a description using wave functions. In contrast to the traditional way of dealing with wave functions, which was superseded by the method of second quantization, Feynman's method allows him to treat the Dirac equation in much the same way that he treated Schrödinger's equation, where second quantization was never claimed to be necessary:

> By a suitable choice and interpretation of the solutions of Dirac's equation the problem may be equally well treated in a manner which is fundamentally no more complicated than Schrödinger's method of dealing with one or more particles. (ThPos, p. 749)

However, Feynman's treatments of Dirac's and Schrödinger's equations, although similar, are not exactly the same. The Green's function for the Dirac equation cannot be expressed in exactly the same way in terms of the eigenfunctions and eigenvalues of the equation's Hamiltonian as for the Hamiltonian in the Schrödinger equation. The problem is[7] that this type of Green's function would describe electrons scattered into negative energy states, which are hard to interpret physically. However, the Green's function for the Schrödinger equation (see (5.5)) is not the only function that satisfies the definition of the Green's function for the Dirac case.[8] A Green's function that would "give results equivalent to those of the positron hole theory" is given by a sum similar to the standard form (5.5) but with summing

[6] ThPos, p. 750.

[7] ThPos, p. 752.

[8] ThPos, equation (12).

over only positive energy eigenvalues for $t_2 > t_1$ and over only negative energy eigenvalues for $t_1 > t_2$.[9]

5.3 The Scattering of Probability Waves

The Hamiltonian H of a quantum mechanical system and the corresponding Green's function K are equivalent in the sense that the dynamics of a quantum mechanical system is fully characterized by the specification of either of them. "For some purposes the specification in terms of K is easier to use and visualize",[10] writes Feynman, without, though, being more explicit about its advantages. It is, however, striking that he does not relate the advantage of that description to the interpretation of a perturbative expansion, which is only introduced as a "simple" *example* "to gain greater familiarity with the K function".[11]

For a non-relativistic free particle, the energy eigenvalues E are continuous. They are related to the mass m and the momentum p of the particle by $E = p^2/2m$. Feynman defines the eigenfunctions of a non-relativistic free particle to be $\phi(x) = \exp(ipx)$ (fn. 4), and the sum $\sum_n$ over all eigenvalues and eigenfunctions in equations such as (5.5) is replaced with an integral over the momentum variable p. Therefore, the Green's function for the Schrödinger equation for a free particle is, "as is well known" (fn. 4)

$$K_0(2,1) = \int \exp[-(ip \cdot x_1 - ip \cdot x_2) - ip^2(t_2 - t_1)/2m]\, d^3p(2\pi)^{-3} \tag{5.6}$$

$$= (2\pi i m^{-1}(t_2 - t_1))^{-\frac{1}{2}} \exp(1/2\; im(x_2 - x_1)^2(t_2 - t_1)^{-1}) \tag{5.7}$$

for $t_2 > t_1$ and zero for $t_2 < t_1$.

If the particle is not completely free but acted on by a potential $U(x, t)$ during the infinitesimal time interval $[t_3, t_3 + \Delta t_3]$, the evolution of the wave function from the beginning to the end of that interval is given by

$$\psi(x, t_3 + \Delta t_3) = \exp(-iH\Delta t_3)\psi(x, t_3). \tag{5.8}$$

Since Δt_3 is infinitesimal, the exponential is given by the first term of its perturbative expansion in the time variable. Therefore,

$$\psi(x_3, t_3 + \Delta t_3) = (1 - iH_0\Delta t_3 - iU\Delta t_3)\psi(x, t_3) \tag{5.9}$$

[9] ThPos, equation (17).

[10] ThPos, p. 750.

[11] ThPos, p. 750.

where $H = H_0 + U$, H_0, denoting the Hamiltonian for the free particle $\frac{1}{2m}\sum_i \frac{\partial^2}{x_i^2}$. In the case where the particle is free throughout, the difference in the wave function is, to first order in U,

$$\begin{aligned}\Delta\psi &= (1 - iH_0\Delta t_3 - iU\Delta t_3)\psi(x, t_3) - (1 - iH_0\Delta t_3)\psi(x, t_3) \\ &= -iU\Delta t_3\psi(x, t_3).\end{aligned} \tag{5.10}$$

This quantity represents the effect of the potential. "We shall call [it] the amplitude scattered by the potential (ThPos, p. 751)."

The time t_3 is supposed to lie between t_1 and t_2. Therefore, between t_1 and t_3 and again between $t_3 + \Delta t_3$ and t_2 the particle is free and its wave function at these moments is given by the free Green's function, K_0, and the initial wave function, $\psi(x_1, t_1)$, or the wave function after scattering, $\psi(x_3, t_3 + \Delta t_3)$:

$$\psi(x_3, t_3) = \int K_0(x_3, t_3; x_1, t_1)\psi(x_1, t_1)\, d^3x_1, \tag{5.11}$$

$$\psi(x_2, t_2) = \int K_0(x_2, t_2; x_3, t_3 + \Delta t_3)\psi(x_3, t_3 + \Delta t_3)\, d^3x_3. \tag{5.12}$$

At the end of the process, at time t_2, because of the potential, the wave function is not what it would have been had the particle propagated freely. "The change in the wave function at 2 brought about by the potential is":

$$\Delta\psi(x_2, t_2) = -i\int K_0(2, 3)U(x_3, t_3)K_0(3, 1)\psi(x_1, t_1)\, d^3x_1\, d^3x_3\, \Delta t_3. \tag{5.13}$$

If the potential acts during a finite time interval, one replaces Δt_3 with dt_3 in the above expression and integrates over t_3:

> In the case the potential exists for an extended time, it may be looked upon as a sum of effects from each interval Δt_3 so that the total effect is obtained by integrating over t_3 as well as x_3. (ThPos, p. 751)

For the double integration over space and time, Feynman abbreviates $d^3x_3\, dt_3$ to $d\tau_3$.

The Green's function, $K^{(1)}(2, 1)$, which represents, to first order in the potential, the difference in the time evolution compared with the case of a free particle, can be read off from (5.13) as the weighting function in the integrand:

$$K^{(1)}(2, 1) = -i\int K_0(2, 3)U(x_3, t_3)K_0(3, 1)d\tau_3. \tag{5.14}$$

One must integrate over all space-time. However, since the free Green's function, $K_0(2, 3)$, is zero for $t_2 < t_3$ and $K_0(3, 1)$ is zero for $t_3 < t_1$, the integration is "automatically" (p. 751) restricted to the range between t_1 and t_2.

The Green's function is "the total amplitude for arrival [of the electron] at x_2, t_2 starting from x_1, t_1" (see Section 5.2). In contrast to the case of a free particle, the Green's function for the case where a potential is present is hard to determine exactly. Using the above expression, Feynman can take into account the effect of the potential approximately: he obtains the first-order correction, $K^{(1)}(2, 1)$, which has to be added to the Green's function of a free particle $K_0(2, 1)$. Higher-order terms will yield increasingly better approximations to the exact Green's function for the case where a potential is present:

$$K(2, 1) = K_0(2, 1) + K^{(1)}(2, 1) + K^{(2)}(2, 1) + \dots \tag{5.15}$$

Feynman proposes a physical interpretation for this approximate quantitative solution in Figure 1 of ThPos (see Fig. 5.1) as well as in the text:

> We can understand the result [(5.15), (5.14)] this way. We can imagine that a particle travels as a free particle from point to point, but is scattered by the potential U. Thus the total amplitude for arrival at 2 from 1 can be considered as the sum of the amplitudes for various alternative routes. It may go directly from 1 to 2 (amplitude $K_0(2, 1)$, giving the zero order term in [(5.15)]). Or (see Figure 1(a)) it may go from 1 to 3 (amplitude $K_0(3, 1)$), get scattered there by the potential (scattering amplitude $-iU(3)$ per unit volume and time) and then go from 3 to 2 (amplitude $K_0(2, 3)$). This may occur for any point 3 so that summing over these alternatives gives [(5.14)]. (ThPos, p. 751)

The figure shows that Feynman interprets the processes described by the Green's function in the wave-mechanical tradition of quantum theory: an incoming probability wave is partly scattered off a potential, which produces an additional outgoing wave that superimposes itself on the undisturbed part of the incoming wave.

In addition, Feynman's physical interpretation of the first-order correction immediately leads him to the second-order correction[12]:

> Again, [the particle] may be scattered twice by the potential (Figure 1(b)). It goes from 1 to 3 ($K_0(3, 1)$), gets scattered there ($-iU(3)$) then proceeds to some other point, 4, in space time (amplitude $K_0(4, 3)$) is scattered again ($-iU(4)$) and then proceeds to 2

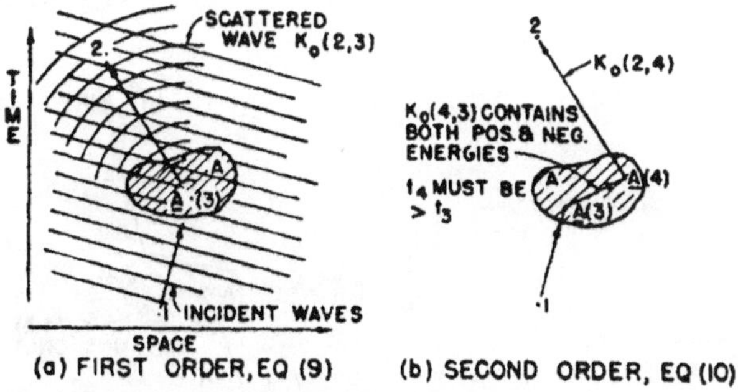

Fig. 5.1 Interpreting Schrödinger's equation. Figure 1 in Feynman's 'Theory of Positrons' (ThPos)

[12] ThPos, p. 751.

($K_0(2,4)$). Summing over all possible places and times for 3, 4 find [sic] that the second order contribution to the total amplitude $K^{(2)}(2,1)$ is

$$(-i)^2 \int\int K_0(2,4)U(4)K_0(4,3)U(3)K_0(3,1)\,d\tau_3\,d\tau_4. \tag{5.16}$$

The possible scattering processes correspond to the possible combinations of Green's functions of free particles in the perturbative expansion of the exact Green's function of a particle acted upon by a potential.

Feynman only hints at the mathematical derivation of higher-order corrections from Schrödinger's equation. And in his view this alternative way of obtaining the quantitative corrections due to a potential to any order justifies the result:

> [(5.16)] can be readily verified directly from [(5.2)] just as [(5.14)] was. One can in this way obviously write down any of the terms of the expansion [(5.15)]. (ThPos, p. 751)

5.4 Waves Propagating Forwards and Backwards in Time

As Feynman writes in the caption to Figure 1 in ThPos, the Dirac equation can also "be visualized as describing the fact that plane waves are scattered successively by a potential". Then, in the caption to Figure 2 (see Fig. 5.2), he points to a peculiarity of the Dirac equation: that the appropriate Green's function is, for negative time intervals, a sum over negative energy states (see the end of Section 5.2, page 116).

To allow for negative time intervals also means introducing a new graphical element into the representation of the processes as the scattering of probability waves. The wave can now not only propagate forwards in time, which is what is usually

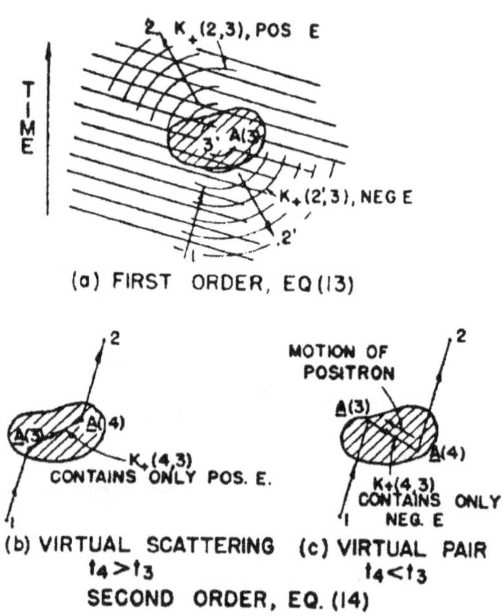

Fig. 5.2 Interpreting Dirac's equation. Figure 2 in ThPos

understood by propagation, but also backwards in time. This is represented by a line pointing towards the direction of negative time. The probability wave scattered towards the negative time direction is interpreted in Fig. 5.2c and in the text as the motion of a positron. This interpretation allows Feynman to use a single mathematical expression to describe processes in which usually two particles are involved. Thus, in his form of representation, processes such as the production of electron–positron pairs are conceived of as the effect of one single electron.

5.4.1 Pairs in Intermediate States

For the Dirac equation, the Green's function for a free particle is denoted by $K_+(2, 1)$ and the perturbation of the free propagation may be brought about by an external electromagnetic potential A. To obtain the perturbative corrections to the free Green's function when an external potential is present, Feynman can simply replace K_0 with K_+ in the corresponding expressions that he found for the Schrödinger equation. Thus, for instance, he obtains for the second-order correction, which has to be added to the Green's function of a free particle $K_+(2, 1)$ in order to give an approximate description of the case where an external potential A is present (cf. (5.16)):

$$-\int\int K_+(2,4)A(4)K_+(4,3)A(3)K_+(3,1)\,d\tau_3\,d\tau_4. \tag{5.17}$$

In the usual positron theory there are two physical processes that contribute quantitatively to such a second-order correction: the electron may be scattered twice by the potential, or an electron–positron pair may be created in an intermediate state of the process. Since the positron is annihilated again, only one electron arrives at the final state, as required. The electron–positron pair is only *virtual*: it does not appear in the initial or final states of the process. Particles that appear in the initial or final states of a process are *real* (see Section 5.4.2).

In Feynman's theory, there seems to be only one second-order correction: (5.17). However, Feynman can show that his one expression in fact includes both contributions, which traditionally are considered separately. Again, Feynman justifies his claims by physically interpreting the equations as describing the successive scattering of a particle. He says:

> That [(5.17)], for example, is the correct second order expression for finding at 2 an electron originally at 1 according to the positron theory may be seen as follows (Figure 2). (ThPos, p. 752)

and then explains his justification. He first observes that (5.17) is valid for $t_2 < t_1$ as well as for $t_2 > t_1$; Feynman chooses to assume the latter case. Furthermore, he assumes, as one would expect, that the potential is only acting in the interval between t_1 and t_2. Then (5.17) is taken to be the total quantitative effect of two processes, the first one being similar to the ordinary scattering of an electron according to Schrödinger's equation considered above. Electrons with negative energy (see end of Section 5.2, p. 116) have no place in Feynman's "reinterpretation" (see the abstract of the article). His formula takes this correctly into account:

> First suppose $t_4 > t_3$ (Figure 2(b)). [t_3 and t_4 are intermediate times between t_1 and t_2.] Then (since $t_3 > t_1$) the electron assumed originally in a positive energy state propagates in that state (by $K_+(3, 1)$) to position 3 where it gets scattered ($A(3)$). It then proceeds to 4, which it must do as a positive energy electron. This is correctly described by [(5.17)] for $K_+(4, 3)$ contains only positive energy components in its expansion, as $t_4 > t_3$. After being scattered at 4 it then proceeds on to 2, again necessarily in a positive energy state, as $t_2 > t_4$. (ThPos, pp. 752–753)

The second process, virtual pair production, which quantitatively contributes to the second-order correction, is also included in the one expression (5.17). The standard theory, in which the positron is viewed as the absence of an electron in a negative energy state, that is, as a "hole", describes the process in the following way:

> In positron theory there is an additional contribution due to the possibility of virtual pair production (Figure 2(c)). A pair could be created by the potential $A(4)$ at 4, the electron of which is that found later at 2. The positron (or rather, the hole) proceeds to 3 where it annihilates the electron which has arrived there from 1. (ThPos, p. 753)

In Feynman's theory, the process of virtual pair production is taken into account if the Green's function, including only negative energy states, is interpreted as the description of the propagation of a positron:

> This alternative [virtual pair production] is already included in [(5.17)] as contributions for which $t_4 < t_3$, and its study will lead us to an interpretation of $K_+(4, 3)$ for $t_4 < t_3$. The factor $K_+(2, 4)$ describes the electron (after the pair production at 4) proceeding from 4 to 2. Likewise $K_+(3, 1)$ represents the electron proceeding from 1 to 3. $K_+(4, 3)$ must therefore represent the propagation of the positron or hole from 4 to 3. (ThPos, p. 753)

This interpretation of the Green's function with negative energy states is, indeed, consistent with Feynman's earlier description of $K_+(4, 3)$: "That it does so is clear," (page 753) he writes, and then explains in detail how $K_+(4, 3)$, for $t_4 < t_3$, describes a hole with *positive* energy, even though $K_+(4, 3)$ only includes *negative* energies in its definition as a sum over energy values E.

Although the Green's function of (5.17) may thus be used to describe two distinct hole theoretical phenomena—the scattering of a single electron and virtual pair production—Feynman proposes an alternative interpretation. An interpretation, that is, in which only one, not two, physical processes corresponds to the one quantitative expression. If a positron is viewed not as a "hole" but as an electron moving backwards in time, the creation of an electron–positron pair and the subsequent annihilation of the positron can be described as the sequence of propagations of a single electron:

> The expressions such as [(5.17)] can still be described as a passage of the electron from 1 to 3 ($K_+(3, 1)$), scattering at 3 by $A(3)$, proceeding to 4 ($K_+(4, 3)$), scattering again, $A(4)$, arriving finally at 2. The scatterings may, however, be toward both future and past times, an electron propagating backwards in time being recognized as a positron. (ThPos, p 753)

5.4.2 The Production and Annihilation of Real Pairs

We have just seen that, by interpreting positrons as electrons "propagating backwards in time", Feynman's description, using the Green's function of the propagation of a single electron, also correctly describes virtual pair production. But this is not all. At the same time, the annihilation of pairs in the initial state of the process and the production of pairs in the final state are also taken into account. Virtual pair production was included in the *second*-order correction to the free propagation of an electron. Real pair production is described by the *first*-order correction term,

$$K_+^{(1)}(2,1) = -i \int K_+(2,3)A(x_3,t_3)K_+(3,1)d\tau_3, \tag{5.18}$$

supposing the interpretation of a positron as an electron propagating backwards in time.

As in the expression for the second-order correction, no restriction is imposed on the ordering of the times t_1, t_2, t_3. In fact, the different possible orderings each describe a different type of scattering and also real pair production and annihilation:

> With this interpretation real pair production is also described correctly (see Figure 3 [Fig. 5.3]). For example in [(5.18)] if $t_1 < t_3 < t_2$ the equation gives the amplitude that if at time t_1 one electron is present at 1, then at time t_2 just one electron will be present (having been scattered at 3) and it will be at 2. On the other hand if t_2 is less than t_3, for example, if $t_2 = t_1 < t_3$, the same expression gives the amplitude that a pair, electron at 1, positron at 2 will annihilate at 3, and subsequently no particles will be present. Likewise if t_2 and t_1 exceed t_3 we have (minus) the amplitude for finding a single pair, electron at 2, positron at 1 created by $A(3)$ from a vacuum. If $t_1 > t_3 > t_2$, [(5.18)] describes the scattering of a positron. (ThPos, p. 753)

The figure (see Fig. 5.3) that accompanies the interpretation given in the text is more abstract than the two preceding figures (see Figs. 5.1 and 5.2). The area where the potential acts on the particles is no longer represented by a contour of the same form, such as the supposed actual contour of the potential in space and time; it is now

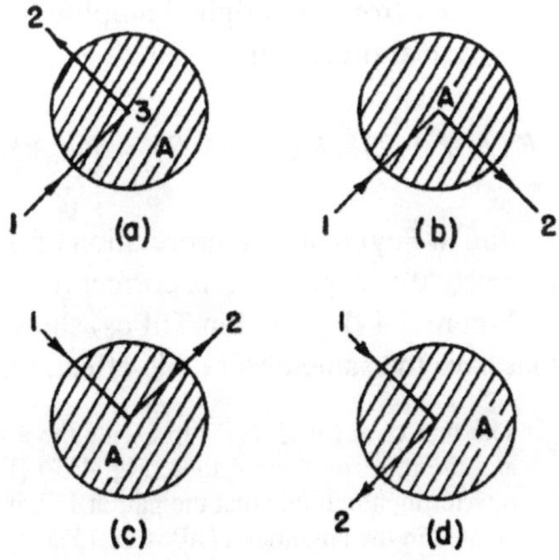

Fig. 5.3 The annihilation (**b**) and production (**c**) of electron–positron pairs represented as special cases of the propagation and scattering off the potential of a single electron. Figure 3 in ThPos

only indicated by a hatched disc. The pattern of lines representing the incoming and outgoing waves has been omitted. The lines perpendicular to the wave front have been retained, although they now indicate the propagation of the particles and not necessarily a wave vector. The figure shows the four possible initial and final states, before and after interaction with a potential, which are, in Feynman's representation, an effect of the propagation of a single electron.

5.5 Generalization to Two Particles

In Feynman's description, even real pair production/annihilation is not, strictly speaking, a two-particle process. Rather, real pair production/annihilation is described as the propagation of one single particle, although the propagation may be backwards in time. In his next step, Feynman generalizes his description to situations with two (charged) particles. However, he still does not take into account the interaction between the charges.

5.5.1 Automatic Inclusion of the Exclusion Principle

When there is no interaction, "each particle behaves independently of the other" (ThPos, p 755) and the probability amplitude of their propagation therefore factorizes. For instance,

$$K(3, 4; 1, 2) = K_{+a}(3, 1)K_{+b}(4, 2) \tag{5.19}$$

is the amplitude of propagation of particle a from 1 to 3 and of particle b from 2 to 4.

However, attention has to be paid to the exclusion principle, which states that no two electrons can be in the same state. In the case of propagation amplitudes, this means that the amplitude with two particles that have interchanged has to be subtracted from the original amplitude and that only the "net amplitude" describes the physical propagation of the two particles, that is (cf. equation 27 in ThPos):

$$K(3, 4; 1, 2) - K(4, 3; 1, 2) = K_{+a}(3, 1)K_{+b}(4, 2) - K_{+a}(4, 1)K_{+b}(3, 2). \tag{5.20}$$

Again, in Feynman's interpretation of positrons as electrons propagating backwards in time, "this expression is correct for positrons also (Figure 4 [Fig. 5.4], p. 755)".

Figure 5.4 (Figure 4 in ThPos) shows the different propagations of two electrons that have the same observable effect, that is, the same initial and final states.

> The first term [in (5.19)] represents the amplitude that the electron proceeds from 1 to 3 and the positron from 4 to 2 (Fig. 4(c)) [Fig. 5.4c], while the second term represents the interfering amplitude that the pair at 1, 4 annihilate and what is found at 3,2 is a pair newly created in the potential. (ThPos, p. 755)

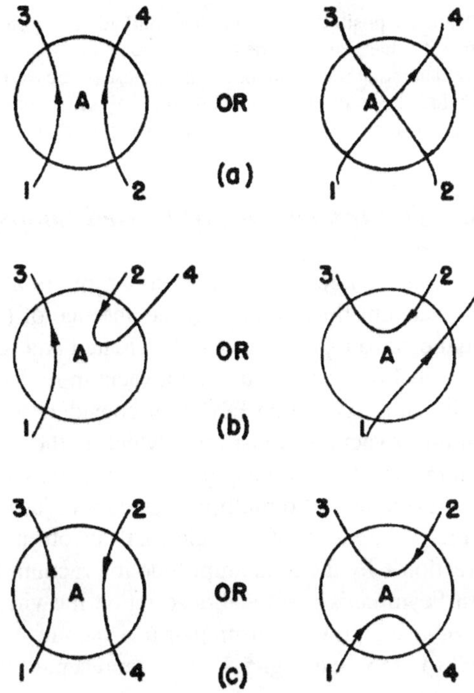

Fig. 5.4 Pairs of processes that differ from each other in the interchange of two electrons. E.g. in (**c**) electrons 3 and 4 are interchanged. Because of the exclusion principle, only the difference of the amplitudes for the two processes is observable. Figure 4 in ThPos

The diagrams articulate the unobservable but nevertheless theoretically important distinction (because of the exclusion principle) of the processes, which differ only in an interchange of the particles. Again, the diagrams are more abstract than the first two figures in ThPos. In Figure 4 the potential takes the form of a circle and not a hatched disc as in Figure 3 of ThPos (see Figs. 5.3 and 5.4). The patterns representing probability waves have again been omitted as they are irrelevant when the exclusion principle is incorporated into Feynman's representation of positrons as electrons propagating backwards in time.

In the standard description, the exclusion principle has further consequences because it also needs to be taken into account in the intermediate states, where virtual pairs may be produced or annihilated. According to Feynman, "no account has to be taken of the exclusion principle in intermediate states" (p. 755) because the exclusion principle, in the case of the two charges considered here, forbids two types of processes. So neither of the two processes is left out in Feynman's description. However, their quantitative contributions to the amplitude cancel each other out. Thus, ignoring the exclusion principle and accordingly including both contributions amount to the same thing as applying the principle and accordingly omitting both terms. Therefore, in Feynman's description, unlike in the standard description, an irrelevant complication is avoided:

> As an example consider again expression [(5.17)] for $t_2 > t_1$ and suppose $t_4 < t_3$ so that the situation represented (Figure 2(c)) is that a pair is made at 4 with the electron proceeding to

2, and the positron to 3 where it annihilates the electron arriving from 1. It may be objected[13] that if it happens that the electron created at 4 is in the same state as the one coming from 1, then the process cannot occur because of the exclusion principle and we should not have included it in our term [(5.17)]. We shall see, however, that considering the exclusion principle also requires another change which reinstates the quantity. (ThPos, p. 755)

5.5.2 The Irrelevancy of Closed Loops

The other change to which Feynman refers at the end of the previous quote concerns vacuum fluctuations. In the absence of (real) electrons or positrons, pairs of virtual electron–positrons can be created and re-absorbed by the external potential. The probability amplitude that a vacuum remains a vacuum is, therefore, not unity. All the amplitudes that Feynman considers are relative to this total amplitude for vacuum-to-vacuum transitions, which is the sum of all amplitudes for vacuum-to-vacuum transitions via any possible intermediate state. To obtain absolute probabilities, Feynman has to multiply the amplitudes he finds—for example, the amplitude for the propagation of a single (real) electron, to which (5.17) is the second-order correction—by the total amplitude for vacuum-to-vacuum transitions.

In Feynman's graphical description the virtual creation and subsequent absorption of an electron–positron pair in a vacuum-to-vacuum transition is a "closed loop path" (p. 755; see Fig. 5.5). It is represented as the cyclic propagation of a single electron.

When calculating the vacuum-to-vacuum amplitude, the exclusion principle again applies. Suppose that we are interested in finding the probability amplitude for the propagation of a single real electron. The exclusion principle dictates that, where the (virtual) electron in the loop is in the same state as the real electron, the contribution has to be omitted. In Feynman's procedure this term is not excluded but it is of exactly the same quantity (with the opposite sign) as the term in the second-order correction for the propagation of a single electron, so they cancel each other out:

> Now one process that can be visualized as occurring in the vacuum is the creation of a pair at 4 followed by a re-annihilation of the *same* pair at 3 (a process which we shall call a closed loop path). But if a real electron is present in a certain state 1, those pairs for which the electron was created in state 1 in the vacuum must now be excluded. We must therefore

Fig. 5.5 A closed loop path representing the creation and subsequent annihilation of a virtual electron–positron pair, in other words, the cyclic propagation of a single electron. Diagram by A. W.

[13] An objection was actually made, probably by Edward Teller, at the conference held at Pocono Manor, Pennsylvania, in 1948 (Weiner 1966, p. 35).

> subtract from our relative amplitude the term corresponding to this process. But this just reinstates the quantity which it was argued should not have been included in [(5.17)] [...] It is obviously simpler to disregard the exclusion principle completely in the intermediate states. (ThPos, p 755)

However, the justification Feynman gives in his paper for disregarding the exclusion principle is somewhat obscure: he does not explicitly specify the terms that would have to be left out by the principle. The specification of the terms is indeed difficult in Feynman's description, since he hardly uses wave functions, which describe general electron states. Feynman's whole presentation applies to states that are characterized by the position of a particle at a particular time. The Green's function describes the transition from one eigenstate of position to another. This restriction to particular states makes it difficult to comprehend how the exclusion principle applies in detail.

Maybe it is clear to Feynman that the two terms forbidden by the exclusion principle will cancel each other out, because in the particular model of the processes he is developing the two processes actually coincide. The graphical representation of the terms (not included in the publication) shows this: if the electron virtually created at 4 (cf. Fig. 5.2, Feynman's sub-figure (c)) is in the same state as the electron coming from 1, then points 1, 2, 3 and 4 come to lie on a straight line. As a result, the N-shaped line of electron propagation becomes a straight line going from 1 to 2 with a superimposed closed loop (see Fig. 5.6). If, on the other hand, the closed loop describing the polarization of the vacuum that can occur independently of the propagation of the electron from 1 to 2 is such that the electron of the vacuum fluctuation is in the same state as the propagating electron, the closed loop comes to lie on the straight line again, which leads to the same diagram. Within a sign the two diagrams correspond to the same amplitude. Thus the amplitudes corresponding to the two diagrams cancel each other out.

Feynman recalls that he did have difficulties treating the virtual creation and absorption of pairs in the vacuum-to-vacuum transition, the so-called *polarization*

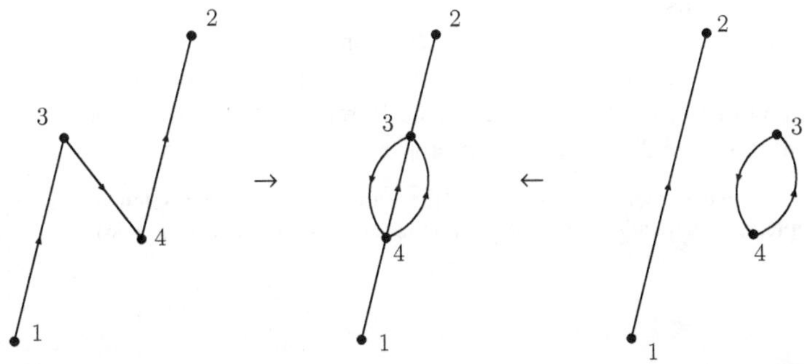

Fig. 5.6 Graphical proof of the identity of the two terms forbidden by the exclusion principle. Diagrams by A. W.

of the vacuum.[14] Feynman computes "all amplitudes relative to the amplitude that a vacuum at t_1 will still be a vacuum at t_2". In a draft of ThPos, in which he discusses the problem of the exclusion principle in intermediate states, he lets us understand that vacuum polarization is only considered in the conventional, and not in his, description (see Fig. 5.7, number 1)[15]:

> According to present theory state T and S might be the same in 2nd term. Usual theory says no because then at time between t_y, t_x can't have 2 electrons in same state. We say it is same electron so Pauli exclusion doesn't operate. Old theory has such a term anyway, for it contemplates pairs created by x anhillated [annihilated] by y (vac. polarization type)—one of which is excluded if electron is in state S—namely the pair created at x (& destroyed at y) whose electron is at S. So term is subtracted relative to (infinite) vacuum if electron is at S. This is same term as we have so both theories give same result here.

It seems that Feynman is trying to find arguments to discount closed loops altogether. In his model, the closed loops are disconnected from the observable physical process and should, therefore, be irrelevant. Indeed, in his next paper, 'Space-Time Approach to Quantum Electrodynamics' (STQED), Feynman does mention (albeit in a footnote) that in his description closed loops do not have to be taken into account:

> There are loops completely without external interactions. For example, a pair is created virtually along with a photon. Next they annihilate, absorbing this photon. Such loops are disregarded on the grounds that they do not interact with anything and are thereby completely unobservable. Any indirect effects they may have via the exclusion principle have already been included. (STQED, fn. 22, p. 780)

Even before Feynman's ThPos and STQED papers are published in September 1949, Freeman J. Dyson proves (in a paper that appeared in February this year), by way of a quantum field theoretic treatment based on Schwinger's theory, that the terms corresponding to closed loops in a graphical representation à la Feynman can be left out (RadTh, see Section 6.4).

Vacuum polarization is one of the sources of the divergence difficulties in the conventional theory. In Feynman's graphical description, these processes are represented as closed loops, which are not connected to the part of the diagrammatic representation of the process in which one is actually interested (the propagation of one electron, for instance). His representation demonstrates that vacuum polarization is a process that does not affect the relevant features of the processes involving electrons in the initial and final states. It only affects the amplitudes as an overall constant. Therefore, Feynman computes all amplitudes relative to the vacuum and thus circumvents the divergence problems:

> Put in this form the theory is complete and there are no divergence problems. Real processes are completely independent of what goes on in the vacuum. (ThPos, p. 756)

---

[14] Weiner 1966, p. 32.

[15] *Theory of Positrons; Self-Energy in an Atom; Radiation Scattering; Spreading Dirac Packet*, folio 6; cf. Schweber 1994, p. 430f.

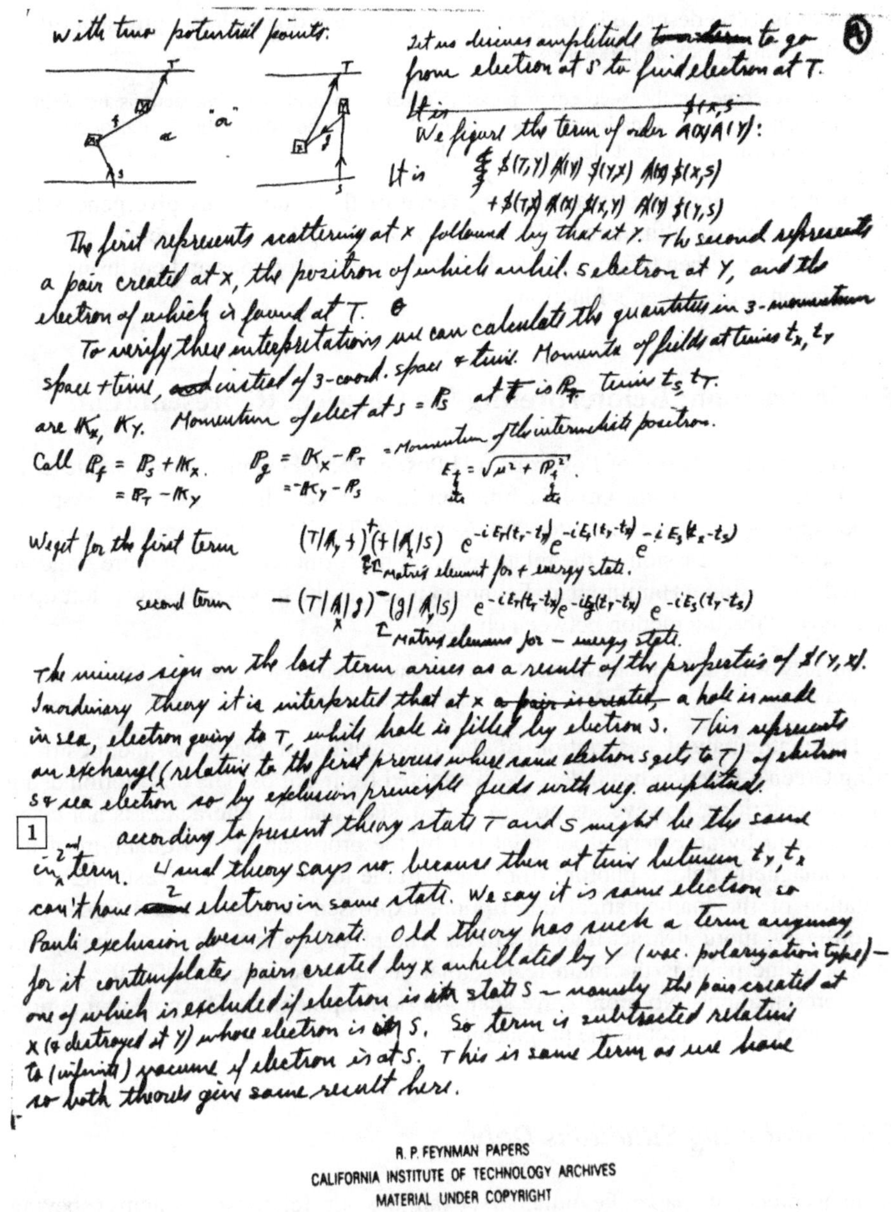
with two potential points.

Let us discuss amplitude ~~to return~~ to go from electron at S to find electron at T.

~~It is~~ $(x,s)

We figure the term of order A(X)A(Y):

It is ∫∫ $(T,Y) A(Y) $(Y,X) A(X) $(X,S) + $(T,X) A(X) $(X,Y) A(Y) $(Y,S)

The first represents scattering at X followed by that at Y. The second represents a pair created at X, the positron of which annihil. S electron at Y, and the electron of which is found at T.

To verify these interpretations we can calculate the quantities in 3-momentum space + time, ~~and~~ instead of 3-coord. space + time. Momenta of fields at times t_X, t_Y are K_X, K_Y. Momentum of elect at S = P_S at T is P_T times $t_S\ t_T$.

Call $P_f = P_S + K_X = P_T - K_Y$ $P_g = K_X - P_T = -K_Y - P_S$ = Momentum of the intermediate positron. $E_f = \sqrt{\mu^2 + P_f^2}$

Weight for the first term $(T|A_Y+)(+|A_X|S)\ e^{-iE_T(t_T-t_Y)} e^{-iE_f(t_Y-t_X)} e^{-iE_S(t_X-t_S)}$ matrix element for + energy state.

second term $-(T|A_X|g)^-(g|A_Y|S)\ e^{-iE_T(t_T-t_X)} e^{-iE_g(t_Y-t_X)} e^{-iE_S(t_Y-t_S)}$ matrix elements for − energy state.

The minus sign on the last term arises as a result of the properties of $(Y,X). In ordinary theory it is interpreted that at X ~~a pair is created,~~ a hole is made in sea, electron going to T, while hole is filled by electron S. This represents an exchange (relative to the first process where same electron S gets to T) of electron S & sea electron so by exclusion principle finds with neg. amplitude.

1 according to present theory state T and S might be the same in 2nd term. Usual theory says no, because then at time between t_Y, t_X can't have 2 electrons in same state. We say it is same electron so Pauli exclusion doesn't operate. Old theory has such a term anyway, for it contemplate pairs created by X annihilated by Y (vac. polarization type) — one of which is excluded if electron is in state S — namely the pair created at X (& destroyed at Y) whose electron is in S. So term is subtracted relative to (infinite) vacuum if electron is at S. This is same term as we have so both theories give same result here.

Fig. 5.7 Notes (probably a draft) to ThPos, ca. 1948 (*Theory of Positrons; Self-Energy in an Atom; Radiation Scattering; Spreading Dirac Packet*, folio 6, reprinted with permission of Melanie Jackson Agency, LLC)

We need to bear in mind that Feynman has not yet considered the interaction that takes place between the particles and, therefore, has not yet had to deal with the divergences arising in these instances. When the interaction of electrons and positrons is to be described, then vacuum polarization does make a physical difference in certain types of processes:

> When we come, in the succeeding paper [STQED], to deal with interactions between charges, however, the situation is not so simple. There is the possibility that virtual electrons in the vacuum may interact electromagnetically with the real electrons. (ThPos, p. 756)

In the electromagnetic interaction lurk some of those notorious divergences that Feynman hopes to eliminate in his alternative description of the theory. The next important step is then to incorporate the interaction of electrons and positrons in his formulation using Green's functions.

5.6 Interaction: Reinterpreting the Classical Representation

As promised in 'Theory of Positrons' (ThPos, p. 755), Feynman treats the description of interaction using Green's function in a "succeeding paper": the 'Space-Time Approach to Quantum Electrodynamics' (STQED). After some introductory remarks and a discussion of the relationship of his formulation to the more standard formulation using a Hamiltonian, Feynman takes up the problem that was left open in ThPos—"the interaction between charges":

> We study by the same methods as in [ThPos], the interaction of two particles using the same notation as [ThPos]. (STQED, p. 771)

The mathematical description of the propagation of electrons and positrons using Green's function has indeed been adopted from ThPos. The description using Green's functions, however, is now extended, such that the interaction is not taken into account by an external potential but by the propagation of a quantum of the electromagnetic field, a photon, from one particle to the other. The resulting interpretation of the mathematical description, expressed in the text and figures, is, accordingly, more abstract than in ThPos. The propagation from space-time point to space-time point is the main feature that needs to be articulated in the graphical representations. No more wave scattering off a potential. The potential is now represented as an effect of the propagation of a photon.

5.6.1 Scattering Situations Only

As in his preceding paper, Feynman first examines non-relativistic systems (obeying Schrödinger's equation) and recalls that in the case of no interaction the Green's function for a system of two particles simply factorizes into the product of the Green's functions of each individual particle (cf. (5.19)). The case becomes significantly more complicated when interaction is involved.

Here again Feynman meets one of the problems that had made him abandon the model of the quivering electron (see Section 4.7.2): he only had found a way to define the wave function for two particles, "for paths which are long enough that they go beyond the time of interaction (this stinks)". The problem does not go away when he moves from trying to justify a Green's function by providing an adequate description of the electron's zigzag paths to the more abstract level where Green's functions are fundamental. Feynman does recognize, though, that this complication is not reason enough to give up the formulation of QED using Green's functions:

> When the particles do interact, one can only define the quantity $K(3, 4; 1, 2)$ precisely if the interaction vanishes between t_1 and t_2 and also between t_3 and t_4. In a real physical system such is not the case. There is such an enormous advantage, however, to the concept that we shall continue to use it, imagining that we can neglect the effect of interactions between t_1 and t_2 and between t_3 and t_4. (STQED, p. 771)

Feynman has to restrict his description to situations where the interactions in the initial and final states are negligible. Typically, these are scattering situations, where two essentially free particles approach one another to interact and then separate, ending up as free particles again. Although the standard formulation of QED is able to describe systems interacting at any time, such as the nucleus of an atom and the electron bound to it, the restriction is reasonable, since, according to Feynman, any physical problem can be treated in this way. The price to pay for doing so without applying the restriction is simply too high:

> Inasmuch as any physical problem can be defined in terms of scattering processes we do not lose much in a general theoretical sense by this approximation. [...] It is essentially to avoid this approximation that the complicated structure of the older quantum electrodynamics has been built up. (STQED, pp. 771–772)

Although, in footnote 9 of the same paper, Feynman retracts one of the assumptions of the theory of classical electrodynamics that Wheeler and he developed,[16] he still wants to quantize a theory of delayed electromagnetic interactions. And although he was not able to include the quantization of such a theory in his thesis, now he has found an appropriate description of it:

> We wish to describe electrodynamics as a delayed interaction between particles. If we can make the approximation of assuming a meaning to $K(3, 4; 1, 2)$ the results of this interaction can be expressed very simply. (STQED, p. 772)

5.6.2 Adapting the Classical Expression for Electrostatic Interaction

In ThPos, Feynman found the first-order correction to the Green's function of an electron described by the Schrödinger equation, perturbed by the external potential, U, during an infinitesimal time interval Δt_0 to be (cf. (5.14)):

$$K^{(1)}(2, 1) = -i \int K_0(2, 3)U(x_3, t_3)K_0(3, 1)\, dx_3\, \Delta t_0. \tag{5.21}$$

---

[16] Cf. Sauer (2008, p. 8).

In STQED, Feynman applies this formula to the electrostatic interaction between two particles, which is no longer described by an external potential. Feynman takes for the potential U the classical expression for the Coulomb potential, e^2/r, where r is the distance of the two particles at a given moment in time and e their charges. In fact, as we will see below, Feynman uses the distance $r_{56} = x_5 - x_6$ between the first particle at time t_5 and the second particle at time t_6, but then restricts the integration to the times where $t_5 = t_6$ through the introduction of a δ function, $\delta(t_5 - t_6)$. Thus the potential now depends on two space-coordinates, x_5 and x_6, instead of on only x_3 as U does in the case of one particle acted upon by an external potential, which he considered in ThPos. As in the one-particle case, one has to integrate over the space-coordinates on which the potential depends.

In the above formula the Green's functions describe the propagation of one electron. $K^{(1)}(2, 1)$ is the first-order correction, which has to be added to the Green's function of a free particle propagating from 2 to 1 when a potential is present. $K_0(3, 1)$ describes the free propagation from the initial space-time point (x_1, t_1) to the space-time point where a unit interaction occurs, (x_3, t_3). Similarly, $K_0(2, 3)$ describes the free propagation from the space-time point where a unit interaction occurs, (x_3, t_3), to the final space-time point (x_2, t_2). Now that two particles are involved, these single-electron expressions have to be replaced by a description of the propagation of two particles. The first-order correction to the propagation of one particle from 1 to 3 and of the other from 2 to 4 is $K^{(1)}(3, 4; 1, 2)$. The free propagation of one particle from its initial space-time point 1 to the space-time point 5, where it interacts with the potential produced by the other particle, and the free propagation of the other particle from its initial space-time point 2 to the space-time point 6, where it interacts with the potential produced by the first particle, is $K_0(5, 6; 1, 2)$. Similarly, $K_0(3, 4; 5, 6)$ describes the free propagation of the two particles after interaction. Finally, the Green's function for two non-interacting particles factorize, such that $K_0(5, 6; 1, 2) = K_0(5, 1)K_0(6, 2)$ and $K_0(3, 4; 5, 6) = K_0(3, 5)K_0(4, 6)$.

From this "obvious generalization to two particles" (STQED, p. 772) Feynman feels justified to claim that the first-order contribution of the correction due to a Coulomb potential mediating the interaction between the particles during the infinitesimal time interval Δt_0 is:

$$K^{(1)}(3, 4; 2, 1) = -ie^2 \iint K_0(3, 5)K_0(4, 6)r_{56}^{-1} K_0(5, 1)K_0(6, 2)\, dx_5\, dx_6\, \Delta t_0. \tag{5.22}$$

If the potential is non-zero during a finite time, one also has to replace "Δt_0" with an integration over t_0.

The Coulomb potential describes an instantaneous interaction, so $t_0 = t_5 = t_6$. The integration over t_0 can, accordingly, be replaced with an integration over t_5 and t_6, weighted by a function $\delta(t_5 - t_6)$, making the integrand non-zero only for $t_5 = t_6$. Thus for the "first order effect of interaction" Feynman obtains:

$$K^{(1)}(3,4;1,2) = -ie^2 \iint K_0(3,5)K_0(4,6)r_{56}^{-1}\delta(t_{56})K_0(5,1)K_0(6,2)\,d\tau_5\,d\tau_6, \tag{5.23}$$

where $d\tau = dx\,dt$ and $t_{56} = t_5 - t_6$.

The Coulomb interaction in classical (relativistic) electrodynamics is the first term in the multipole expansion of the exact interaction potential described by Maxwell's equations. This description by a Coulomb potential has the particular drawback that it describes an instantaneous interaction, which is not well-defined relativistically. The relativistic interaction between two particles is delayed by the time that it takes the electromagnetic potential to travel—at the speed of light c—from where it is produced to where it acts on a particle. For particles a distance r_{56} away, the potential needs the time r_{56}/c to travel from one particle to the other. Feynman sets $c = 1$, such that the delay is simply r_{56} in appropriate units.

The only element in the above formula that contains the information that the interaction is instantaneous is the δ function, which restricts the integration to the domain where $t_5 = t_6$. If the potential is delayed by r_{56}, those integrands contribute, for which $t_5 = t_6 \pm r_{56}$. Feynman first considers the case where $t_6 = t_5 - r_{56} < t_5$. The delay of the interaction by r_{56} "suggests simply replacing $r_{56}^{-1}\delta(t_{56})$ in [5.23] by something like $r_{56}^{-1}\delta(t_{56} - r_{56})$ to represent the delay in effect of b [second particle] on a [first particle]" (STQED, p. 772).

5.6.3 Interpreting the Potential as the Propagation of Photons

A well-known feature of any electromagnetic field is that it can be treated as an assembly of harmonic oscillators.[17] This is evident from the so-called Fourier transformation of the field. Feynman mentions this feature in an earlier article:

> By taking the Fourier transform of [the field produced by one of the particles], one can represent the fields as a superposition of the effects of harmonic oscillators in the usual way. (CutOffCl, p. 942)

In that article, Feynman then proposes to "cut-off" the function that describes which oscillator frequencies contribute to the interaction in order to arrive at a divergence-free electrodynamics—classical as well as quantum.[18]

In the quantum case, harmonic oscillators can take on discrete energy values, each energy leap corresponding to the energy of a quantum of the electromagnetic field. Feynman understands the interaction to be the effect of the "emission

[17] See, e. g., Fermi 1932, Part III.

[18] CutOffCl; CutOffQ.

and absorption of a virtual quantum" or the effect of "quanta which can be exchanged".[19]

Interaction by the "exchange" of a quantum is a concept that had already been in use for some time, in particular in the case of nuclear forces.[20] The problem with this notion in the field of electrodynamics lay in the fact that the electrostatic (Coulomb) interaction requires something like *longitudinally* polarized photons rather than observable *transversely* polarized photons. In a type of move that we saw him take with the exclusion principle, Feynman recognizes that the difficulties can be avoided simply by, as it were, not paying too much attention to them. For instance, on how to calculate the self-energy of the electron, Feynman comments:

> The treatment of the longitudinal self-energy is usually different, for the longitudinal oscillators are first eliminated from the Hamiltonian, their effect being the term e^2/r_{00} where r_{00} is the meaningless distance of the electron from itself. [...] Fortunately, all these points may be most easily circumvented by simply not eliminating the longitudinal oscillators from the field Hamiltonian at all. (CutOffQ, p. 1431)

Thus Feynman is more prepared than other physicists to adopt the concept of photon exchange as the mechanism of electromagnetic interaction. While working on the cut-off papers (CutOffCl; CutOffQ), Feynman becomes used to representing a classical electromagnetic potential as an assembly of harmonic oscillators and, in the quantum case, to conceiving of the interaction as being brought about by emissions or absorptions of the quanta of these oscillators—the photons. However, when he applies this representation to the expression he has obtained in terms of Green's functions for the delayed version of the Coulomb interaction (5.23), he encounters a problem:

> This turns out not to be quite right,[footnote] for when this interaction is represented by photons they must be of only positive energy, while the Fourier transform of $\delta(t_{56} - r_{56})$ contains frequencies of both signs. (STQED, p. 772)

In order to have quanta of only positive frequency, Feynman restricts the integration in the Fourier transform of $\delta(t_{56} - r_{56})$ to positive values of the frequency ω, instead of all real values. The Fourier transform of $\delta(x)$ is $\mathrm{e}^{-i\omega x}/\pi$, which gives, for the modified δ function,

$$\delta_+(x) = \int_0^\infty \mathrm{e}^{-i\omega x}\, d\omega/\pi, \tag{5.24}$$

which describes a delayed version of the electrostatic interaction by positive-frequency photons.

So far only the effect of particle b on a has been considered, that is, the case where $t_5 > t_6$. For the opposite order of times, the appropriate interaction term is $r_{56}^{-1}\delta_+(-t_{56} - r_{56})$. Neither of the possibilities $t_5 > t_6$ and $t_5 < t_6$ is preferred,

---

[19] CutOffQ, pp. 1434, 1437.

[20] See, e. g., Carson (1996a, b).

both contributing equally to the integral in (5.23). Therefore, the terms from the two possibilities have to be "averaged" (p. 772) in order to obtain the joint effect of interaction. The averaging process gives another δ function, one with the relativistic space-time interval between the two points of interaction as an argument. The division by distance r disappears:

$$\frac{1}{2}\left(r_{56}^{-1}\delta_+(t_{56}-r_{56})+r_{56}^{-1}\delta_+(-t_{56}-r_{56})\right)=\delta_+\left(t_{56}^2-r_{56}^2\right). \tag{5.25}$$

As a result, to describe a delayed version of the electrostatic interaction, Feynman replaces $r_{56}^{-1}\delta(t_5-t_6)$, describing the ordinary instantaneous electrostatic interaction, with $\delta_+(s_{56}^2)$, where $s_{56}^2=t_{56}^2-r_{56}^2$ is the relativistic invariant space-time interval between points (x_5,t_5) and (x_6,t_6).

Besides the electrostatic interaction described by the *scalar* potential, there is also the interaction via the *vector* potential. In order to consider the vector potential, the term $-(v_5\cdot v_6)\delta_+(s_{56}^2)$ needs to be added to $\delta(s_{56}^2)$, taking into account the scalar potential. In the relativistic quantum case described by the Dirac equation, the velocities v_5 and v_6 correspond to the operators α_a and α_b. The joint effect of the interaction via the scalar and vector potentials of two electrons described by Dirac's equation is, therefore,

$$(1-\alpha_a\alpha_b)\delta_+\left(s_{56}^2\right), \tag{5.26}$$

which, by using the relationships between the Dirac operators, can be written as:

$$\beta_a\beta_b\gamma_{a\mu}\gamma_{b\mu}\delta_+\left(s_{56}^2\right), \tag{5.27}$$

(see STQED, p. 772, between equations 3 and 4).

When, moreover, one replaces the Green's function for an electron described by Schrödinger's equation with the Green's function K_+ for an electron described by Dirac's equation and absorbs the β operators in the definition of K_+, Feynman finally arrives at his "fundamental equation for electrodynamics" (p. 772):

$$K^{(1)}(3,4;1,2)=-ie^2\iint K_+(3,5)K_+(4,6)\gamma_{a\mu}\gamma_{b\mu} \times\delta_+\left(s_{56}^2\right)K_+(5,1)K_+(6,2)\,d\tau_5\,d\tau_6. \tag{5.28}$$

This is equation (4) in STQED.

Feynman's equation is "a consequence of conventional electrodynamics" (p. 772) but his interpretation of it and the status he attributes to it are not so conventional:

> It describes the effect of exchange of one quantum (therefore first order in e^2) between two electrons. It will serve as a prototype enabling us to write down the corresponding quantities involving the exchange of two or more quanta between two electrons or the interaction of an electron with itself. (STQED, p. 772)

Feynman, thereby, reduces all electrodynamics to one single process, any process being representable as a combination of that fundamental process.

The fundamental process of QED consists of two electrons and a photon the free propagation of which is interrupted by the emission or absorption of the photon by one of the electrons. Feynman does not deduce that the term "$\delta_+(s_{56}^2)$" describes the free propagation of a photon in the same way that he has deduced that K represents the propagation of an electron. The function δ_+ is, first of all, not the Green's function associated with the photon wave equation as the Ks are with the Dirac or Schrödinger equations. The δ_+ function is introduced as a term to represent the potential energy between the two electrons. The interpretation of the δ_+ function is determined by the interpretation of the other elements of the fundamental equation.

Only in a footnote (no. 8) of Feynman's paper (STQED), in which he explains how to treat real (rather than virtual) photons, does Feynman mention that the δ_+ function satisfies the photon wave equation with a δ function (without the "+"!) as the inhomogeneity and is, therefore, by definition, a Green's function of that equation (see Section 5.1):

$$-\Box_2^2 \delta_+\left(s_{21}^2\right) = 4\pi\delta(2, 1), \tag{5.29}$$

where

$$-\Box_2^2 = \frac{\partial^2}{\partial t_2} - \frac{\partial^2}{\partial x_2^{(1)}} - \frac{\partial^2}{\partial x_2^{(2)}} - \frac{\partial^2}{\partial x_2^{(3)}},$$

$x_2^{(i)}$ being the ith component of the space coordinate x_2.[21]

It is possible that it is Feynman's correspondence with Welton that draws his attention to this relationship. In his letter, Welton represents the Klein–Gordon equation in an integral form, which he interprets as being contributions starting from "the last scattering" (see Section 4.2). The photon wave equation is simply the special case of the Klein–Gordon equation for a particle with zero mass. In the same letter Welton also suggests that, in order to treat the Dirac equation in a similar way, "we need a generalization of the $\delta(s^2)$ function". Feynman's δ_+ might, in part, be the result of the search for that generalization.

5.7 The Right "picture"

Feynman's interpretation implies that any quantum electrodynamic process can be described as the composition of one basic process. The processes can be analyzed as a more or less complex composition of the fundamental process. Quantitatively, the compositions are described by the perturbative expansion of the Green's function

[21] Feynman has introduced this notation in his previous paper, ThPos, p. 757.

(cf. (5.15)). If a process is analyzed as a more complex composition, the quantitative predictions become more precise.

In order to establish the role of his "fundamental equation", which describes the basic process of QED, Feynman goes on to give a more detailed interpretation of the "exchange of one quantum between two electrons":

> We shall now interpret Eq. [5.28] in a manner which will permit us to write down the higher order terms. It can be understood (see Fig. 1 [Fig. 5.8]) as saying that the amplitude for "a" to go from 1 to 3 and "b" from 2 to 4 is altered to first order because they can exchange a quantum. Thus, "a" can go to 5 (amplitude $K_+(5,1)$) emit a quantum (longitudinal, transverse, or scalar $\gamma_{a\mu}$) and then proceed to 3 ($K_+(3,5)$). Meantime "b" goes to 6 ($K_+(6,2)$), absorbs the quantum ($\gamma_{b\mu}$) and proceeds to 4 ($K_+(4,6)$). The quantum meanwhile proceeds from 5 to 6, which it does with amplitude $\delta_+(s_{56}^2)$. We must sum over all the possible quantum polarizations μ and positions and times of emission 5, and of absorption 6. Actually if $t_5 > t_6$ it would be better to say that "a" absorbs and "b" emits but no attention need be paid to these matters, as all such alternatives are automatically contained in [5.28]. (STQED, pp. 772–773)

The interpretation given in the text is complemented by a diagram that could arguably be classified as being the first *Feynman diagram*. However, as I will discuss in Chapter 6, the Feynman diagrams that became established are, in fact, Dyson's modified versions of them. The diagram here is tied to the context of wave mechanics, while the Feynman diagrams modified by Dyson are interpreted in the context of the field theoretical second quantization. The Feynman diagrams we use today are understood either in the context of second quantization or in the context of a modern version of path-integral quantization.

If one compares the diagrams in ThPos with the diagrams in STQED, one sees that Feynman has introduced a new graphical element into the first Feynman diagram: the wavy line which represents the propagation of the photon that brings about the interaction. The wavy line is simply a modification of the lines that indicate the propagation of the electromagnetic potential in the diagrams used in classical theories of electrodynamics (see Figs. 4.29 and 4.30).

Feynman's interpretation of his fundamental equation is not one of classical trajectories of particles bouncing off each other. First of all, the forces described in terms of the fundamental process of the emission or absorption of a photon by an electron or a positron might also be *attractive*. This would be impossible in an

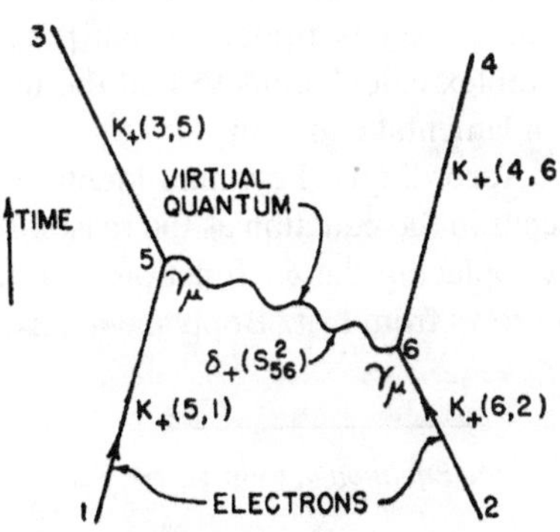

Fig. 5.8 Diagram that Feynman uses to interpret his "fundamental equation". Figure 1 in STQED

interpretation akin to the scattering of billiard balls, where there is always recoil and thus only repulsive forces. Feynman's interpretation is much more abstract. The Green's functions K describe the propagation of a particle but not its trajectory in the classical sense. Even though in the graphical representation K is represented as a straight line, which might suggest a trajectory, it is rather a sum of contributions from all possible trajectories (see Section 5.2).

Feynman's interpretation is also abstract because, unlike in classical mechanics, the sequence of emission and absorption events is not definite. However, this is not a shortcoming but an advantage of the interpretation, which recognizes that the sequence of intermediate events is irrelevant. As Feynman says in the passage quoted above: "no attention need be paid to these matters, as all such alternatives are automatically contained in (5.28)."

With this fundamental equation, Feynman's "struggle" to understand the Dirac equation comes to an end. He has now found the form of representation—the "picture" in Feynman's words—a "slight modification" of which would "straighten out some of the present troubles".[22] At various points in the 'Space-Time Approach to Quantum Electrodynamics', Feynman emphasizes that his fundamental equation is equivalent to the standard formulation of QED, at least as far as quantitative predictions are concerned. Therefore, Feynman's formulation inherits all the diseases, as it were, from the standard formulation: quantitative predictions for observable quantities such as the energy of a single electron and a certain type of spectroscopic line shifts are infinite and thus at variance in the most extreme possible way with experience and in fact completely uninterpretable as an observable physical quantity:

> These expressions are, as has been indicated, no more than a re-expression of conventional quantum electrodynamics. As a consequence, many of them are meaningless. For example, the self-energy expression [equation numbers] gives an infinite result when evaluated. The infinity arises, apparently, from the coincidence of the δ function singularities in $K_+(4, 3)$ and $\delta_+(s_{43}^2)$. Only at this point is it necessary to make a real departure from conventional electrodynamics, a departure other than simply rewriting expressions in a simpler form. (STQED, p. 776)

However, Feynman has now reduced all QED to a fundamental equation that describes the fundamental process from which all electrodynamic phenomena are made up. If the whole of electrodynamics is essentially an exchange of photons between electrons, then the uninterpretable divergences must have their roots in an inadequate description of that process. No need, then, to search through pages of complex calculations to find the infinities. All that one has to do is investigate the fundamental equation.

Accordingly, Feynman identifies the singularities in the Green's functions that occur in the equation as the reason for the divergences. He can thus treat the disease by replacing the δ_+ function with the smoother function f, an idea that Feynman borrows from Fritz Bopp and a thesis written by H. McManus under the supervision

---

[22] *Dirac Equation a*, folio 12, see Section 4.4.

of Rudolf Peierls.[23] Feynman had already applied this idea in an earlier paper (CutOffCl) in such a way that the resulting modified classical theory of electrodynamics might be quantized (see Section 5.6.3). Of the quantization, however, Feynman only published "some preliminary results" (CutOffCl, p. 939), since he still did not have a "complete picture" (CutOffQ, p. 1430). The "complete physical theory may be published in the near future", he wrote (CutOffQ, p. 1430). In STQED, he then tries to keep his word:

> We desire to make a modification of quantum electrodynamics analogous to the modification of classical electrodynamics described in a previous article [CutOffCl]. There the $\delta\left(s_{12}^2\right)$ appearing in the action of interaction was replaced by $f\left(s_{12}^2\right)$ where $f(x)$ is a function of small width and great height. (STQED, p. 776)

Feynman's references in earlier papers such as CutOffCl to later (yet to be published) results indicate that the development of his ideas is not necessarily reflected in the order of publication. It is, therefore, hard to determine whether Feynman *first* arrives at his fundamental equation and identifies the δ_+ function as the cause for all the problems and then finds a replacement in the cut-off papers; or that, knowing of a way to remove the divergences arising from the δ functions in classical electrodynamics, he tries to mould QED into a formulation that will exhibit these types of divergences.

---

[23] Bopp 1940, 1942, cf. McManus 1948; see footnote 4 of CutOffCl.

Chapter 6
The Field Theoretical Systematization of Feynman's Theory

6.1 Widespread Misunderstanding: The Need for Systematization

Feynman gave the first public presentation of his alternative formulation of quantum electrodynamics (QED) at a "conference on physics" that was sponsored by the National Academy of Sciences and held at Pocono Manor, Pennsylvania, United States, from 30 March to 1 April 1948.[1] It proved to be a complete disaster! The audience failed to understand how Feynman could possibly justify his results, even if it was agreed that the results were correct.

Julian Schwinger gave the preceding presentation, which, despite the fact that it lasted several hours, was largely appreciated on account of Schwinger's mathematical deductions. However, when it came to Schwinger giving a physical interpretation of his formulae, "the wolves were out, and it was terrible".[2] Consequently, Hans Bethe advised Feynman to restrict his presentation to the mathematical deductions of his results from the principles of his theory rather than to try to justify those principles by physically interpreting them. Feynman recalls Bethe saying[3]:

> You'd better explain this thing mathematically and not physically because look how much trouble Schwinger has every time he says anything physical. (Weiner 1966, p. 34)

Feynman, however, had most probably intended to present his fundamental equation ((5.28), see Sections 5.6.3 and 5.7) and its interpretation as the fundamental process of the exchange of a photon between two electrons. Instead, he found himself having to deduce his results from a relativistic version of his path integral action.[4] This version differed from the one attempted at the end of RMP48 (see Section 3.5) and would be published only 3 years later.[5] At the time of the conference he had

---

[1] Wheeler (1948), see also Schweber (1994), Chapter 4 and Chapter 8 (section 10), p. 436ff.

[2] Weiner (1966, p. 33); see also Schweber (1994, p. 436).

[3] All quotes from Weiner (1966) are reprinted with permission. Copyright 1966, American Institute of Physics

[4] Weiner (1966, p. 34); Wheeler (1948), in particular p. 47; Schweber (1994), in particular p. 440.

[5] Feynman (1951).

A. Wüthrich, *The Genesis of Feynman Diagrams*, Archimedes 26,
DOI 10.1007/978-90-481-9228-1_6, © Springer Science+Business Media B.V. 2010

probably not even worked out all the details. In any case, only at short notice did Feynman decide to present his results as the consequence of an action principle and path integrals.

Bethe's intentions were surely honourable, but the conference participants failed to appreciate the presentation. They wanted a justification of the action principle and were not impressed by Feynman's single fact that his theory gave the right predictions. Feynman remembers conversations like the following[6]:

Feynman: This is a mathematical formula which I will now show you produces all the results of quantum electrodynamics.
Audience: Where does the formula come from?
Feynman: Never mind where the formula came from, it works. It's the right formula.
Audience: How do you know it's right?
Feynman: Because it gives all the right answers.

Even when Feynman showed, by way of examples, that correct results could, indeed, be deduced from his formula, his audience demanded a physical justification for it. This implied that Feynman would have to explain his alternative formulation by way of path integrals, which would not have been necessary had he started with the less idiosyncratic solution to Schrödinger's and Dirac's equations by Green's functions that led to his fundamental equation. Feynman recollects:

> "What made you think the formula was right in the first place?" [asked the audience]. So then I had to go with physical ideas. And then I was much deeper in the soup, because I'd come at them in the wrong order, and everything was chaotic.
> I started to explain about path integrals and all this stuff. [. . .][7]

Here Feynman recalls that his physical justification was not appreciated since it was given *after* he had presented the action principle. I find it implausible that a good justification would lose its power to convince only because it was presented in the wrong order. Rather, I take this episode as confirmation that Feynman's initial intention had been to present his fundamental process of photon exchange described by Green's functions as the principle of his theory.

During Feynman's exposition of his formulation of QED by path integrals, the criticisms did not cease. Edward Teller, for one, was concerned about the exclusion principle (cf. Section 5.5)[8]:

Teller: What about the exclusion principle?
Feynman: It doesn't make any difference in intermediate states.
Teller: How do you know?
Feynman: I know, I worked from a—
Teller: How could it be?

---

[6] Weiner (1966, p. 34).

[7] Weiner (1966, pp. 34–35).

[8] Weiner (1966, p. 35).

Paul Dirac worried about whether the total probability of all alternative processes would add up to unity[9]:

Dirac: Is it unitary?
Feynman: Is *what* unitary?
Dirac: The matrix which carries you from the present position to the future position.
Feynman: I haven't got any matrix which carries me from the present position to the future position, I go forwards and backwards and forwards in time. So I don't know.

And Niels Bohr was troubled by whether Feynman's notion of paths was consistent with the uncertainty principle. As Feynman recalls:

> "Already in 19 something"—1924, '25, or something—"we know that the classical idea of the trajectory in a path is not a legitimate idea in quantum mechanics" and so on. In other words he [Bohr] [is] telling me about the uncertainty principle, you see, and so on. (Weiner 1966, p. 36)

All these criticisms, in particular Bohr's, were, to a large extent, due to the fact that the audience had misunderstood Feynman's theory. This might not have been the case had Feynman been able to start his presentation with his fundamental process and its description by Green's functions. These functions would surely have been familiar to most of the audience, while the physical interpretation of his fundamental equation would certainly not have sounded absurd, since it involved familiar processes of emission and absorption of quanta. The issues raised during Feynman's presentation can only to a small extent be seen as genuine criticisms of Feynman's theory. Referring to Bohr's criticism, Feynman recalls:

> I mean, it was no trouble. It's just that he wasn't understanding at all. And I simply got a feeling of resignation. It's very simple, I'll have to publish this and so on, let them read it and study it, because it's right. (Weiner 1966, p. 36)

Later, at the same meeting, Bohr would apologize to Feynman, and Bohr's son, Aage, who was also present, would tell Feynman that his father had indeed misunderstood parts of Feynman's exposition.[10]

6.2 Dyson Intervenes

In February 1949, before, in September, Feynman's STQED and ThPos are published, a paper by Freeman J. Dyson appears (RadTh) in which he compares Feynman's theory to attempts by Schwingerand Sin-Itiro Tomonaga to remove

---

9 Weiner (1966, p. 35).

10 See Schweber (1994, p. 444).

QED's uninterpretable divergences.[11] To eliminate the infinities they propose a method based on the second quantized theory of conventional QED.[12]

Although Feynman has not yet published his theory by the time that Dyson's paper appears, Dyson is already considerably knowledgeable about Feynman's theory.[13] Dyson was a visiting graduate student at Cornell University (Ithaca, NY), where the young Feynman was a member of staff, and the two became close friends for a time. And Dyson, unlike Feynman, was also familiar with the more conventional method of second quantization (see also end of Section 4.5). Thus Dyson was able to mediate between Feynman and the community of theoretical physicists, which was, in general, familiar with second quantization. Feynman recollects:

> So Dyson was there and he knew what I was doing. But at that time, I had not proved in any way [the] relationship [of Feynman's technique] to the standard quantum electrodynamics—in other words, what was wanted by many people would have been this[:] Start with the standard formulation of quantum electrodynamics, with operators, creation and annihilation operators and all kinds of theoretic things, and you show that as a consequence of that, these rules and these formulas of mine were right. Then you could be perfectly satisfied. There was only one trouble. The author of the formulas, namely myself didn't know anything about creation and annihilation operators and all this correct formulation of electrodynamics, because he never learned it. (Weiner 1966, p. 41)

In his paper, Dyson proves the equivalence of the three approaches taken by Feynman, Schwinger and Tomonaga, which enables him to use Feynman's less complicated procedures to obtain the predictions but then justify the results using the firmer theoretical foundation of Tomonaga's and Schwinger's theory. Feynman considers such a justification to be superfluous and at best of pedagogical interest to those who were, like most physicists, familiar with second quantization. He recollects:

> He [Feynman] was therefore considerably too impatient to learn all that stuff which he'd done all this work to avoid, you see, in order to prove to people who happened to have learned that, that they don't need it, that the rules are easier, and I felt very strongly that everything was simpler than the regular formulation, because even Schwinger was working things out and would write a whole lot of stuff before he got to a point that I could write down immediately with my diagrams. (Weiner 1966, p. 41)

6.2.1 The Most Appropriate State Vector

Basically the solution to all the problems in QED amounts to finding the solutions to a Schrödinger equation describing the time evolution of a state vector Φ. The state vector contains all the information one can have about the system. The dynamics of

---

[11] Appendix B contains a chronological overview of the relevant publications.

[12] See, e. g., Schwinger (1948b); Tomonaga (1946).

[13] See, e. g., Schweber (1994, p. 502ff.) and Gleick (1992), sections "Cross-country with Freeman Dyson" and "Oppenheimer's surrender". See also Kaiser (2005), section "The First Apprentice", pp. 65–74.

the system is described by the Hamiltonian of the system. The Schrödinger equation relates the two elements as follows:

$$i\hbar[\partial/\partial t]\Phi = \left\{\int H(r)\,dr\right\}\Phi, \tag{6.1}$$

where $H(r)$ is the total energy density of the system.[14] The general solution to this equation is given by an integral involving the Hamiltonian "with Φ_0 any constant state vector"[15]:

$$\Phi(t) = \exp\left\{[-it/\hbar]\int H(r)\,dr\right\}\Phi_0. \tag{6.2}$$

In the non-relativistic Schrödinger equation, which is a differential equation for the wave function, the derivative is with respect to time. For a relativistic theory, it is more appropriate to use a derivative with respect to a space-like surface $\sigma(r)$. The "Tomonaga–Schwinger form of the Schrödinger equation" gives us this form (RadTh, p. 487):

$$i\hbar c[\partial\Phi/\partial\sigma(r)] = H(r)\Phi. \tag{6.3}$$

Again, the equation can be solved by multiplying a constant state vector with an integral that involves the Hamiltonian containing the information on the system's dynamics. The definition of a "constant" state vector, however, must also be consistent with relativistic principles. A constant state vector is, therefore, not one for which "the results of measurements of field quantities at any given point in space are independent of time" (pp. 487–488). This would be too strong a condition. Rather, one should only stipulate that a constant state vector describes "a system [consisting] of photons, electrons, and positrons, traveling freely through space without interaction or external disturbance" (p. 488). To formulate such a definition, Dyson divides the total Hamiltonian into a part H_0, which describes the energy-density of free electrons and electromagnetic fields, and a part H_1, which describes the interaction between the electrons and the fields:

$$H(r) = H_0(r) + H_1(r). \tag{6.4}$$

If a constant state vector Ψ_0 is defined to meet the aforementioned requirements, the time evolution of a general state vector Ψ depends only on the interaction Hamiltonian H_1. The Schrödinger equation for Ψ thus reads:

$$i\hbar c[\partial\Psi/\partial\sigma(r)] = H_1(r)\Psi. \tag{6.5}$$

---

[14] RadTh has $d\tau$ instead of dr, the infinitesimal spatial volume element. I assume this to be a misprint.

[15] RadTh, p. 487.

The Schrödinger equation for Ψ can be solved using a perturbative expansion:

$$\Psi(\sigma) = U(\sigma_0)\Psi_0, \tag{6.6}$$

with

$$U(\sigma_0) = 1 + (-i/\hbar c)\int_{-\infty}^{\sigma_0} H_1(x_1)\,dx_1 + (-i/\hbar c)^2 \times \int_{-\infty}^{\sigma_0} dx_1 \int_{-\infty}^{\sigma(x_1)} H_1(x_1)H_1(x_2)\,dx_2 + \dots. \tag{6.7}$$

Ψ_0 is still not the most appropriate vector to describe the constant state of a system in the context of electrons, positrons and their interaction with the electromagnetic field. This is because there are two relevant types of interaction, namely the interaction between the particles through an electromagnetic field that they themselves generate and the effect of an *external* electromagnetic field on the particles. To take into account this distinction, Dyson introduces a further division of the energy density. This time he divides the interaction energy-density H_1 into an internal and an external part:

$$H_1(x_0) = H^{\mathrm{i}}(x_0) + H^{\mathrm{e}}(x_0). \tag{6.8}$$

The internal part, H^{i}, is given by the electromagnetic current j_μ, which corresponds to the configuration of the electrons and positrons in the system, and the vector potential A_μ, which corresponds to the configuration of the photons in the system:

$$H^{\mathrm{i}}(x_0) = -[1/c]j_\mu(x_0)A_\mu(x_0). \tag{6.9}$$

The external part, H^{e}, describes the external forces, in most cases an external electromagnetic force described by the potential A_μ^{e} (superscript "e" for "external").

For "the most important systems, [those] in which only one particle is actually present" (p. 489), the most appropriate definition of a constant state vector is a vector that describes a system devoid of any external electromagnetic influences. Such a system is not constant in the strict sense of the word because an electron, say, will virtually create photons as described by H^{i}. Although any observation will always detect a single (real) electron, a system that contains a single electron at all times does not exist. It is, therefore, appropriate to define the constant vector Ω_0 as one that includes the effect of H^{i}. The dynamics of a general vector Ω is then determined by the influence of the external electromagnetic field alone. The Schrödinger equation for Ω thus reads:

$$i\hbar c[\partial\Omega/\partial\sigma(x_0)] = (S(\sigma))^{-1}H^{\mathrm{e}}(x_0)S(\sigma)\Omega. \tag{6.10}$$

The operator S is the same as U, with H^{i} replacing H_1.

By using $S(\sigma))^{-1}H^{\mathrm{e}}(x_0)S(\sigma)$ instead of $H(r)$ in the Schrödinger equation, the internal interaction or "radiation interaction" (RadTh, p. 489) is incorporated into the state vector and no longer appears as a factor causing a change in the system's state. This definition of a state vector is the most appropriate for systems consisting of a single particle because their internal interactions are not observable. Such a state vector describes the interaction of an observable, or *physical*, electron with an external electromagnetic potential.

For general systems, however, interpreting the state vector Ω is more subtle:

> The elimination of the radiation interaction is hereby achieved; only the question, "How is the new state vector $\Omega(\sigma)$ to be interpreted?", remains.
>
> It is clear from [(6.10)] that a system with a constant Ω is a system of electrons, positrons, and photons, moving under the influence of their mutual interactions, but in the absence of external fields. In a system where two or more particles are actually present, their interactions alone will, in general, cause real transitions and scattering processes to occur. For such a system, it is rather "unphysical" to represent a state of motion including the effects of the interactions by a constant state vector; hence, for such a system the new representation has no simple interpretation. However, the most important systems are those in which only one particle is actually present, and its interaction with the vacuum fields gives rise only to virtual processes. In this case the particle, including the effects of all its interactions with the vacuum, appears to move as a free particle in the absence of external fields, and it is eminently reasonable to represent such a state of motion by a constant state vector. Therefore, it may be said that the operator,
>
> $$H_{\mathrm{T}}(x_0) = (S(\sigma))^{-1}H^{\mathrm{e}}(x_0)S(\sigma), \tag{6.11}$$
>
> on the right of [(6.10)] represents the interaction of a physical particle with an external field, including radiative corrections. Equation [(6.10)] describes the extent to which the motion of a single physical particle deviates, in the external field, from the motion represented by a constant state-vector, i. e., from the motion of an observed "free" particle. (RadTh, pp. 489–490)

However, this definition of state vector is still problematic. The energy-density H^{i} includes the energy of the electron in its own field and is, therefore, infinite. Because of that, the operator S has an infinite expectation value in any state: $\langle S\rangle_\Omega = (\Omega^*, S\Omega) = \infty$. However, for a constant state vector Ω_0, it is $S(\infty)\Omega_0 = \Omega_0$, from which the expectation value $\langle S\rangle_{\Omega_0} = 1$ follows. There is, therefore, no Ω_0 that could satisfy the definition of a constant state vector.[16]

This inconsistency can be eliminated by distinguishing between the "bare" mass of an electron, which it would have if no internal radiative interactions occurred, and the mass of an observable electron. The inertia of a "bare" electron would be determined by the mass parameter m implicitly contained in H_0. Such a "bare" electron, however, does not exist, since internal radiative interactions are inevitable. The electron always produces virtual electromagnetic fields, positrons and other electrons, which modify the inertia of the electron; an *electromagnetic* mass has to be added to the "bare" mass to describe an observable electron:

---

16 See RadTh, p. 490.

> The mistake that has been made occurred in trying to represent the observed electron with its electromagnetic self-energy by a wave-field with the same characteristic rest-mass as that of the "bare" electron. To correct the mistake, let δm denote the electromagnetic mass of the electron, i. e., the difference in rest-mass between an observed and a "bare" electron. Instead of [(6.4)], the division of the energy-density $H(r)$ should have taken the form
>
> $$H(r) = (H_0(r) + \delta m\, c^2 \psi^*(r)\beta\psi(r)) + (H_1(r) - \delta m\, c^2 \psi^*(r)\beta\psi(r)). \tag{6.12}$$
>
> The first bracket on the right here represents the energy-density of the free electromagnetic and electron fields with the observed electron rest-mass [...] (RadTh, p. 490)

The definition of Ω, the state vector that already includes internal radiative interactions, should, therefore, use

$$H^{\mathrm{I}}(x_0) = H^{\mathrm{i}}(x_0) + H^S(x_0) = H^{\mathrm{i}}(x_0) - \delta m\, c^2 \bar{\psi}(r)\psi(r), \tag{6.13}$$

instead of H^{i} alone ($\bar{\psi}(r) = \psi^*(r)\beta$).

With (6.10) and the definition and interpretation of the state vector Ω, Dyson has prepared the main ingredients for his presentation of Schwinger's and Feynman's theories in the remainder of the article:

> The important results of the present paper up to this point are [equation (6.10)] and the interpretation of the state vector Ω. The [ordinary quantum mechanical] state vector Ψ of a system can be interpreted as a wave function giving the probability amplitude of finding any particular set of occupation numbers for the various possible states of free electrons, positrons, and photons. The state vector Ω of a system with a given Ψ on a given surface σ is, crudely speaking, the Ψ which the system would have had in the infinite past if it had arrived at the given Ψ on σ under the influence of the interaction $H^{\mathrm{I}}(x_0)$ alone. (RadTh, p. 490)

If there is no external force acting on the system, its state vector Ω remains constant; it does not change from the infinite past to any other moment σ in space-time: $\Omega(-\infty) = \Omega(\sigma)$. The state vector Ψ ordinarily used to describe quantum mechanical systems, however, would change, even in the absence of external forces. The ordinary description would ascribe to the system a different state vector at σ than at $-\infty$. Dyson's alternative description ascribes the state $\Psi(-\infty)$ to the system for all moments in space-time when no external forces are present.

Dyson's definition and interpretation of Ω take the infinite past as reference for a constant state vector. Therefore, his description singles out the time-direction towards the future. To restore the symmetry between the past and the future, Dyson defines Ω'. The difference, however, is not of great relevance:

> The definition of Ω being unsymmetrical between past and future, a new type of state vector Ω' can be defined by reversing the direction of time in the definition of Ω. Thus the Ω' of a system with a given Ψ on a given σ is the Ψ which the system would reach in the infinite future if it continued to move under the influence of $H^{\mathrm{I}}(x_0)$ alone. More simply, Ω' can be defined by the equation
>
> $$\Omega'(\sigma) = S(\infty)\Omega(\sigma). \tag{6.14}$$

Since $S(\infty)$ is a unitary operator independent of σ, the state vectors Ω and Ω' are really only the same vector in two different representations or coordinate systems. Moreover, for any steady state the two are identical [...] (RadTh, pp. 490–491)

6.2.2 Green's Functions Interpreted as Vacuum Expectation Values

After having defined a suitable state vector and introduced the electromagnetic contribution to the mass of the electron, Dyson is ready to propose the "fundamental formulas of the Schwinger and Feynman theories" (section V of RadTh). According to Dyson these are the power series of the Hamiltonian operator H_T, the "effective external potential energy" (p. 491), (see (6.11) and (6.10)). The Hamiltonian H_T can be expressed in terms of the Hamiltonian H^{I}, which describes the interactions between the electrons, positrons and photons of the system, and the H^{e}, which describes the effect of the external forces on the system.

In Schwinger's theory (or, rather, Tomonaga's; hence the subscript "T" is added to the H), the said expression follows from (6.11) and the definition of $S(\sigma)$ and reads

$$H_{\mathrm{T}}(x_0) = \sum_{n=0}^{\infty}(i/\hbar c)^n \int_{-\infty}^{\sigma(x_0)} dx_1 \int_{-\infty}^{\sigma(x_1)} dx_2 \ldots \times \int_{-\infty}^{\sigma(x_{n-1})} dx_n[H^{\mathrm{I}}(x_n), [\ldots, [H^{\mathrm{I}}(x_2), [H^{\mathrm{I}}(x_1), H^{\mathrm{e}}(x_0)]] \ldots]]. \tag{6.15}$$

Dyson mentions that, using the above power series, Schwinger succeeded in calculating, for example, the magnetic moment of the electron, the results of which were confirmed by experimental evidence.[17] The drawback, however, was the complexity of the formula:

The repeated commutators in this formula are characteristic of the Schwinger theory, and their evaluation gives rise to long and rather difficult analysis. (RadTh, p. 491)

To obtain predictions from Schwinger's theory, one has to calculate the matrix elements of "H_{T} between states specified by their state vectors Ω" (p. 491), $\Omega_2^* H_{\mathrm{T}} \Omega_1$. According to Dyson, in Feynman's theory one uses a "mixed representation" (p. 491): an Ω' and an Ω to specify the matrix elements of the Hamiltonian (subscript "F" for Feynman's version of the Hamiltonian):

$$\Omega_2'^* H_{\mathrm{F}} \Omega_1. \tag{6.16}$$

[17] Schwinger (1948a); RadTh, p. 491.

Because of the relationship

$$\Omega_2'^{*} = \Omega_2^{*} S^{*}(\infty), \tag{6.17}$$

which follows from equation (6.14), Feynman's matrix elements are equivalent to the matrix element

$$\Omega_2^{*} S^{*}(\infty) H_{\mathrm{F}} \Omega_1 \tag{6.18}$$

between two Ωs (no prime!). Feynman's version of the Hamiltonian H_{F} and Schwinger's version are, therefore, related by

$$H_{\mathrm{T}}(x_0) = S^{*}(\infty) H_{\mathrm{F}}(x_0), \tag{6.19}$$

that is, Feynman's Hamiltonian is (S is unitary, therefore $SS^{*} = 1$)

$$H_{\mathrm{F}} = S(\infty) H_{\mathrm{T}}(x_0) = S(\infty)(S(\sigma))^{-1} H^{\mathrm{e}}(x_0) S(\sigma). \tag{6.20}$$

Feynman's fundamental formula, according to Dyson, is the power series of H_{F}:

$$H_{\mathrm{F}}(x_0) = \sum_{n=0}^{\infty} (i/\hbar c)^n [1/n!] \int_{-\infty}^{\infty} dx_1 \ldots \int_{-\infty}^{\infty} dx_n \times P(H^{\mathrm{e}}(x_0), H^{\mathrm{I}}(x_1), \ldots, H^{\mathrm{I}}(x_n)). \tag{6.21}$$

P abbreviates the prescription of applying the operators $H^{\mathrm{e}}(x)$ and $H^{\mathrm{I}}(x)$, not necessarily in the order in which they appear in the above expression but in the order determined by the time component of their argument.

Dyson did not intend to provide a close rendition of Feynman's theory through the above formula. Rather, it is a version of the Schwinger–Tomonaga theory that incorporates Feynman's equal treatment of positive and negative time directions:

> In the Feynman theory the basic principle is to preserve symmetry between past and future. Therefore, the matrix elements of the operator H_T are evaluated in a "mixed representation". (RadTh, p. 491)

For further information on Feynman's theory, Dyson refers the reader to RMP48, CutOffCl, CutOffQ and Wheeler and Feynman 1945 at the beginning of his article (in footnote 3) and comments: "These articles describe early stages in the development of Feynman's theory, little of which is yet published."

In Dyson's paper, "the Schwinger theory will be carried no further" (p. 491). Dyson's aim is to derive Feynman's results from the quantum field theory of Schwinger (and Tomonaga). If this can be done, the two approaches are equivalent as far as quantitative predictions are concerned. The proof consists of calculating, in the Schwinger–Tomonaga theory, the transition probabilities from one state to another, the aforementioned "matrix elements" of H_F, and showing that, like the mathematical expressions in Feynman's theory, they can be brought into one-to-one correspondence to the diagrams Feynman developed for other reasons.

> The ultimate aim is to obtain a set of rules by which the matrix element of the operator $[H_F]$ between two given states may be written down in a form suitable for numerical evaluation, immediately and automatically. The fact that such a set of rules exists is the basis of the Feynman radiation theory; the derivation in this section of the same rules from what is fundamentally the Tomonaga–Schwinger theory constitutes the proof of equivalence of the two theories. (RadTh, pp. 492–493)

Dyson restricts his proof to transitions from an initial state to a final state with just one electron, "the most important systems" according to him (p. 489), and where the interaction of the one electron with an external electromagnetic field is given by the potential energy

$$H^{\mathrm{e}}(x_0) = -[1/c] j_\mu(x_0) A_\mu^{\mathrm{e}}(x_0). \tag{6.22}$$

$A_\mu^{\mathrm{e}}(x_0)$ is the electromagnetic four-potential and $j_\mu(x_0)$ the charge-current four-vector. The charge and the current are produced by the electrons and positrons as described by the relationship

$$j_\mu(x) = iec\bar{\psi}(x)\gamma_\mu\psi(x) \tag{6.23}$$

between the charge-current four-vector and the electron–positron field operators $\bar{\psi}(x)$ and $\psi(x)$. These are the sums of creation and annihilation operators a and $\bar{a}$ weighted by the spinor wave functions ϕ of free electrons and positrons:

$$\psi_\alpha(x) = \sum_u \phi_{u\alpha}(x) a_u, \tag{6.24}$$

$$\bar{\psi}_\alpha(x) = \sum_u \bar{\phi}_{u\alpha}(x)\bar{a}_u. \tag{6.25}$$

In a similar way the electromagnetic field operator $A_\mu(x)$, which describes the field produced by the particles themselves, is a weighted sum of photon annihilation and creation operators, b_v and $\bar{b}_v$:

$$A_\mu(x) = \sum_v \left(A_{v\mu}(x) b_v + A^*_{v\mu}(x)\bar{b}_v\right). \tag{6.26}$$

The interaction of the particles with that field is taken into account by

$$H^{\mathrm{I}}(x_0) = -[1/c] j_\mu(x_0) A_\mu(x_0). \tag{6.27}$$

$P(H^{\mathrm{e}}(x_0), H^{\mathrm{I}}(x_1), \ldots, H^{\mathrm{I}}(x_n))$ occurring in (6.21) is the sum of products of the field operators, which are themselves the sums of the creation and annihilation operators according to the formulae above. The method Dyson proposes to evaluate the matrix elements of

$$P(H^{\mathrm{e}}(x_0), H^{\mathrm{I}}(x_1), \ldots, H^{\mathrm{I}}(x_n))$$

is to find a criterion according to which one can determine the products contained in the sum that would yield zero as a value for the matrix element and consequently leave them out. The terms giving non-zero matrix elements are then shown to be expressible as the products of the two basic functions, D_{F} and S_{F}, and a factor depending on the wave function of the initial and the final electron.

Let P_n denote the P expression occurring in the nth term of the power series. Inserting the expressions for H^{e} and H^{I} into P_n reveals that it is the sum of products of the operators ψ, $\bar{\psi}$ and A. The As commute with the ψs and $\bar{\psi}$s and can, therefore, be grouped together. The ψs and $\bar{\psi}$s have to be chronologically ordered according to the time-component of their space-time argument x_i. P_n thus contains products of type

$$Q_n = \bar{\psi}(x_{i_0})\psi(x_{i_0})\bar{\psi}(x_{i_1})\psi(x_{i_1})\ldots\bar{\psi}(x_{i_n})\psi(x_{i_n})A(x_{j_1})\ldots A(x_{j_m}). \tag{6.28}$$

In each of the field operators, ψ, $\bar{\psi}$, and A, creation and annihilation operators are contained: the operators in ψ and $\bar{\psi}$ describe the creation and annihilation of electrons and positrons; the operators in A describe the creation and annihilation of photons. Between two states containing exactly one electron, an operator such as Q_n only gives non-zero matrix elements if one annihilation operator annihilates the electron contained in the initial state, one creation operator creates the electron in the final state and all the other creation and annihilation operators are paired in such a way that each created particle is annihilated again. All other Q_ns appearing in the sum P_n can be dispensed with.

In the remaining Q_ns, one can group into pairs the creation and annihilation operators of an electron or a photon and the two operators creating or annihilating the electron in the final or initial states without changing the value of the matrix element, apart from a sign depending on whether the grouping consists of an even or odd number of permutations. Since one member of each pair is a creation operator and the other an annihilation operator, the first is contained in a ψ and the other in a $\bar{\psi}$, or in two As if the particle created and annihilated is a photon. For the two members of a ψ–$\bar{\psi}$ pair, the time-ordering has to be preserved because of the prescription $P(\ldots)$. The non-zero matrix elements are, therefore, included in the matrix elements of all possible products Q_n' of paired ψs, $\bar{\psi}$s and As of Q_n:

$$\begin{aligned} Q_n' &= \varepsilon P(\bar{\psi}(x_0), \psi(x_{r_0}))\ldots P(\bar{\psi}(x_n), \psi(x_{r_n})) \\ &\quad \times P(A(x_{s_1}), A(x_{t_1}))\ldots P(A(x_{s_h}), A(x_{t_h})); \end{aligned} \tag{6.29}$$

$\varepsilon = \pm 1$, depending on whether the permutations that are needed to obtain the pairings are even or odd.

A pair of factors is the sum of products of a creation and an annihilation operator. Only these terms will give a non-zero contribution to the matrix element in which the creation and annihilation operators refer to the same particle. These terms represent the creation and subsequent annihilation of a particle out of the vacuum. (The particles in the initial and final states are annihilated and created by two single factors.) The matrix elements for the paired factors are, therefore, those between two vacuum states. In other words, this is the " 'vacuum expectation value' of the operator and has been calculated by Schwinger" (p. 494). "It turns out" that the vacuum expectation values of the paired factors (ordered by the prescription P) are the Green's functions K_+ and δ_+ that describe electron/positron and photon propagation in Feynman's theory (see Sections 5.4.1 and 5.6.3 (5.24)). Dyson denotes K_+ by S_{F} and δ_+ by D_{F}. Apart from proportionality constants, they equal the time-ordered vacuum expectation values of the field operators:

$$\langle P(\bar{\psi}_\alpha(x), \psi_\beta(y))\rangle_0 = \frac{1}{2}\eta(x, y) S_{\mathrm{F}\beta\alpha}(x - y), \tag{6.30}$$

$$\langle P(A_\mu(x), A_\nu(y))\rangle_0 = \frac{1}{2}\hbar c \delta_{\mu\nu} D_F(x - y). \tag{6.31}$$

The matrix elements that one needs to take into account are, therefore, of the form

$$M = \varepsilon' \prod_{i \neq k} \left(\frac{1}{2} S_{\mathrm{F}}(x_i - x_{r_i})\right) \prod_j \left(\frac{1}{2}\hbar c D_{\mathrm{F}}(x_{s_j} - x_{t_j})\right) \bar{\psi}(x_k)\psi(x_{r_k}) \tag{6.32}$$

with the two single factors, $\bar{\psi}(x_k)$ and $\psi(x_{r_k})$, creating and annihilating the initial and final electron. The matrix elements of these single factors cannot be given in a general form since they differ according to the different wave functions of the electron in the initial and final states. Pairs with $i = i_n$ are not allowed, since in that case the expression (6.23) for the charge-current operator is incorrect. However, with the correct expression the vacuum expectation value for these pairs vanishes. Therefore, one is not making an error by simply leaving them out.

The problem of finding the matrix elements of H_{F} between two one-electron states is thus reduced to evaluating S_{F} and D_{F} and the matrix elements of the single factors. One has to do this for all possible Ms, which are obtained by different choices of pairings and single factors. However, the matrix elements of

$$P(H^{\mathrm{e}}(x_0), H^{\mathrm{I}}(x_1), \ldots, H^{\mathrm{I}}(x_n)),$$

which appear in the expression (6.21) for H_{F}, other than those obtainable by these pairing prescriptions, do not have to be considered, since they are zero. In the end one has to insert all the Ms into the expression (6.21) and perform the summation and integrations. The summation is truncated at a certain value of n, which corresponds to the nth order of approximation in perturbation theory.

This is a general procedure for obtaining the matrix elements of H_{F}, which, according to Dyson, is the essential content of Feynman's theory. However, Dyson's article does not end here:

The problem of calculating the matrix elements of [6.21] is thus in principle solved. However, in the following section it will be shown how this solution in principle can be reduced to a much simpler and more practical procedure. (RadTh, p. 495)

6.3 An Algorithm for Pairing Operators

For the "much simpler and more practical procedure", Dyson introduces prescriptions for translating the mathematical expression M into a symbol containing vertices and two types of lines, such that one M corresponds exactly to one symbol:

> Let an integer n and a product P_n occurring in [6.21] be temporarily fixed. The points $x_0, x_1, \ldots, x_n$ may be represented by $(n+1)$ points drawn on a piece of paper. A type of matrix element M as described in the last section will then be represented graphically as follows. For each associated pair of factors $(\bar{\psi}(x_i), \psi(x_{r_i}))$ with $i \neq k$, draw a line with a direction marked in it from the point x_i to the point x_{r_i}. For the single factors $\bar{\psi}(x_k), \psi(x_{r_k})$, draw directed lines leading out from x_k to the edge of the diagram, and in from the edge of the diagram to x_{r_k}. For each pair of factors $(A(x_{s_i}), A(x_{t_i}))$, draw an undirected line joining the points x_{s_i} and x_{t_i}. The complete set of points and lines will be called the "graph" of M; clearly there is a one-to-one correspondence between types of matrix elements and graphs, and the exclusion of matrix elements with $r_i = i$ for $i \neq k$ corresponds to the exclusion of graphs with lines joining a point to itself. The directed lines in a graph will be called "electron lines", the undirected lines "photon lines". (RadTh, p. 495)

These prescriptions for translating M into a graph serve to find all the non-zero matrix elements M to a given P_n (RadTh, section IX). First

$$P_n = P(H^{\mathrm{e}}(x_0), H^{\mathrm{I}}(x_1), \ldots, H^{\mathrm{I}}(x_n))$$

has to be expanded in terms of ψs, $\bar{\psi}$s and As. Next, "all possible graphs G with $(n+1)$ vertices are now drawn as described [above]". Dyson is not very explicit as to how exactly this should be done. Furthermore, when he considers a concrete example (section X), he only laconically remarks:

> Next, all admissable graphs with the three vertices x_0, x_1, x_2 are to be drawn. It is easy to see that there are only two such graphs, that G shown in Figure 1 [Fig. 6.1], and the identical graph with x_1 and x_2 interchanged. (RadTh, p. 501)

The lack of explications obscures the main function of the diagrams in Dyson's procedure for calculating the matrix elements of H_{F}. A given P_n determines a set of

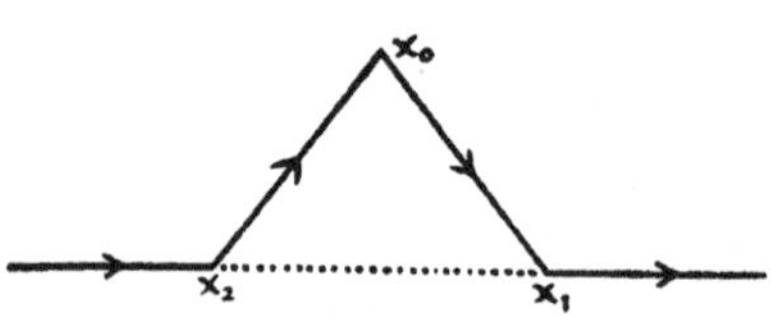

Fig. 6.1 Example of a Feynman diagram, or "graph", which Dyson uses to find the relevant matrix elements of Schwinger's and Tomonaga's theories. Figure 1 of 'The Radiation Theories of Tomonaga, Schwinger, and Feynman' (RadTh, p. 501)

vertices $x_0, \ldots, x_n$. Each x_i is the argument of a ψ, a $\bar{\psi}$ and an A (A^{e} for x_0). Therefore, on the graphical level, each vertex is the starting or end point of two electron lines and one photon line. In the entire graph, one electron line has no starting point (incoming line) and one has no end point (outgoing line). The other electron lines and all the photon lines connect two vertices to each other. The electron lines "form one open polygon" (p. 495) in which the direction of the line is never inverted.

From these rather inexplicit graphical prescriptions, a set of possible graphs is defined. The elements of these graphs can then be translated into mathematical expressions from which one obtains M:

> First, for each photon line joining x and y in G, replace two factors $A_\mu(x)A_\nu(y)$ in P_n (regardless of their positions) by $\left[\left(\frac{1}{2}\hbar c D_F(x-y)\right)\right]$. Second, for each electron line joining x to y in G, replace two factors $\bar{\psi}_\alpha(x)\psi_\beta(y)$ in P_n (regardless of positions) by $\left[\left(\frac{1}{2} S_F(x-y)\right)\right]$. Third, replace the remaining two factors $P(\bar{\psi}_\gamma(z), \psi_\delta(w))$ in P_n by $\bar{\psi}_\gamma(z)\psi_\delta(w)$ in this order. [...] (RadTh, p. 501)

That the mathematical expressions M can be reduced to a diagram, described by Dyson as the "graphical representation of matrix elements", might do away with the need to write down a long integral expression in some cases, but this is not at all the main function of the diagrams. This visualization of the mathematical expressions is secondary. Rather, the most important function of the diagrams is that the graphical rules of connecting vertices and the subsequent translation into a mathematical expression form an algorithm to find all the non-zero matrix elements, and these alone.

Let us now consider the example Dyson gives in the last section of his article. There the "rules will be used for writing down the terms giving second-order radiative corrections to the motion of an electron in an external field" (p. 501). As a second-order correction the summation in (6.21) has to be performed up to $n = 2$. The second-order correction, J_2, is given by a double integral of the matrix elements of P_2:

$$J_2 = ie^3 \int_{-\infty}^{\infty} dx_1 \int_{-\infty}^{\infty} dx_2\, P\left(\bar{\psi}_\alpha(x_0)(\gamma_\lambda)_{\alpha\beta}\psi_\beta(x_0)A^{\mathrm{e}}_\lambda(x_0),\right.$$
$$\bar{\psi}_\gamma(x_1)(\gamma_\mu)_{\gamma\delta}\psi_\delta(x_1)A_\mu(x_1),$$
$$\left.\bar{\psi}_\varepsilon(x_2)(\gamma_\nu)_{\varepsilon\zeta}\psi_\zeta(x_2)A_\nu(x_2)\right). \tag{6.33}$$

There is one possibility for pairing the two non-external electromagnetic potentials $A_\mu(x_1)$ and $A_\nu(x_2)$, and two possibilities for choosing a $\bar{\psi}$ and a ψ as single factors (x_0 cannot be chosen as the argument of a single factor). Thus there are two "admissible graphs" but they only differ in that x_1 and x_2 are exchanged and therefore yield the same quantitative contribution, so we only need to consider one of the graphs and finally multiply the result by 2. By following the translation rules presented above, one obtains the second-order correction in terms of the two single factors $\bar{\psi}(x_1)$ and $\psi(x_2)$, the external potential $A^{\mathrm{e}}_\mu(x_0)$ and the D_{F} and S_{F} functions:

$$L = -i[e^3/8\hbar c] \int_{-\infty}^{\infty} dx_1 \int_{-\infty}^{\infty} dx_2\, D_F(x_1 - x_2) A_\mu^e(x_0) \\ \times \bar{\psi}(x_1)\gamma_\nu S_F(x_0 - x_1)\gamma_\mu S_F(x_2 - x_0)\gamma_\nu \psi(x_2). \quad (6.34)$$

The expression obtained by the procedure involving diagrams of the second-order correction has some advantages over the equivalent expression in Schwinger's theory, but Dyson still finds it hard to evaluate numerically:

> The above expression L is formally simpler than the corresponding expression obtained by Schwinger, but the two are easily seen to be equivalent. In particular, the above expression does not lead to any great reduction in the labor involved in a numerical calculation of the Lamb shift. Its advantage lies rather in the ease with which it can be written down. (RadTh, p. 502)

Dyson tries to establish a fundamental difference in the use of the diagrams in his and in Feynman's procedures:

> In Feynman's theory the graph corresponding to a particular matrix element is regarded, not as merely an aid to calculation, but as a picture of the physical process which gives rise to that matrix element. For example, an electron line joining x_1 to x_2 represents the possible creation of an electron at x_1 and its annihilation at x_2, together with the possible creation of a positron at x_2 and its annihilation at x_1. This interpretation of a graph is obviously consistent with the methods, and in Feynman's hands has been used as the basis for the derivation of most of the results, of the present paper. (RadTh, p. 496)

Dyson claims that the diagrams are above all a means of practicably facilitating numerical results. However, if one looks at the passage quoted before, it is clear that this is only true in a limited sense. The diagrams do not help him to work out the numerical integration required by L, and neither are the D_F and S_F functions, which in turn are defined by an integral, more easily evaluated.

In a statement of Dyson's in a lecture he would give at Cornell University 2 years later, Dyson gets more to the point concerning the function of the diagrams:

> [To calculate M] it is only necessary to pick out from [6.21] the terms which contain the right combination of elementary emission and absorption operators to annihilate the particles in [the initial state] and to create those in [the final state]. We shall next describe a general method of systematically picking out those terms, which is due originally to Feynman. (Dyson 2006, p. 95)

The diagrams help in finding all the non-zero matrix elements in a way that ensures that no possibility is omitted and that no possibility is included twice. The diagrams one can draw starting from P_n connect the vertices x_i in exactly the same way as when the operators are paired when expressing P_n as the sum of vacuum expectation values.

6.4 The Diagrams Reduced to Their Essentials

Translating the mathematical expressions into Feynman diagrams allows Dyson to introduce two further simplifications into his algorithm for calculating matrix elements. The diagrams make this possible because they are an appropriate articulation of the relevant aspects of the phenomena. As such they were constructed by Feynman. In the field theoretical context in which Dyson now places the diagrams, the mathematical justification of the simplifications comes first, and the diagrams become merely an aid for remembering the simplifications.

6.4.1 Disconnected Parts

First, Dyson proves that the *disconnected* parts of a graph can be dispensed with when evaluating the matrix elements of H_F. A disconnected part of a graph is a set of vertices that are connected to each other by at least one line (of any type) but are not connected by any line to any other vertex of the graph. To prove this, Dyson considers the simplest type of a disconnected graph, that is, a graph that consists of two disconnected parts. The part, or subgraph, including the vertex x_0, is the "essential part" (p. 496), since it contains the space-time point where the external potential acts. This subgraph is called G_1. The other subgraph, the "inessential part", is called G_2 (see Fig. 6.2).

Since the matrix element to be calculated can be expressed as a pure product (see (6.32)), the contribution $C(G)$ of the whole graph is the product of the contributions of each subgraph, thus

$$C(G) = C(G_1)C(G_2). \tag{6.35}$$

Dyson proceeds by showing that the contribution of all possible disconnected graphs G with the same G_1 is simply the contribution of G_1. He differentiates between two cases: first, the vertices corresponding to the arguments x_k and x_{r_k} of the two single factors, creating and annihilating the particles in the initial and final states, lie in G_2 (see Fig. 6.2a); second, they lie, instead, in G_1 (Fig. 6.2b). The vertices corresponding to the arguments of the two single factors must lie in the same subgraph; if not, the two subgraphs would not be disconnected.

If the vertices of the two single factors lie in the inessential part G_2 (Fig. 6.2a), G_2 contributes a part of the matrix element of the unit operator between the initial and final states (see RadTh, p. 496). If one sums up the contributions of all possible G_2s, one obtains the complete matrix element of the unit operator between the initial

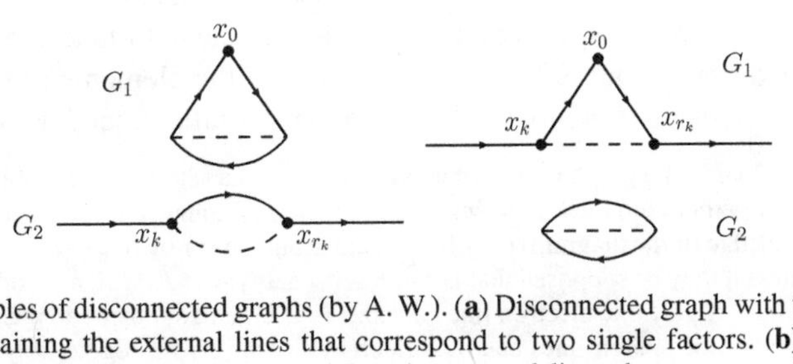

Fig. 6.2 Examples of disconnected graphs (by A. W.). (**a**) Disconnected graph with the "inessential part", G_2, containing the external lines that correspond to two single factors. (**b**) Disconnected graph with the "essential part", G_1, containing the external lines that correspond to two single factors

and final states. However, this matrix element is zero for the cases considered here where a reaction takes place, that is, where the initial and final states are different. Therefore, $C(G)$ is zero for all Gs that have the arguments of the single factor in the inessential part G_2.

For the other graphs, with the single factors in G_1 (Fig. 6.2b), the contribution of all possible G_2s is the expectation value in the vacuum of the unit operator. This is unity and therefore $C(G) = C(G_1)$ for all these graphs. Since in the first case $C(G) = 0$, it is proved that

$$C(G) = C(G_1) \tag{6.36}$$

for all disconnected graphs. That is, only the essential part of a graph has to be considered in quantitative evaluations.

By their graphical elements and the interpretation of these elements, Feynman diagrams show that one part of the diagram is irrelevant:

> The elimination of disconnected graphs is, from a physical point of view, somewhat trivial, since these graphs arise merely from the fact that meaningful physical processes proceed simultaneously with totally irrelevant fluctuations of fields in the vacuum. (RadTh, p. 496)

In Feynman's 'Space-Time Approach to Quantum Electrodynamics' (STQED) the non-inclusion of disconnected graphs only merits a footnote (see Section 5.5). The example that Feynman cites therein is the "inessential part" of Fig. 6.2b.

6.4.2 Self-Energy Parts

Since Dyson has succeeded in translating the mathematical expressions into a more appropriate articulation of the description in terms of Feynman diagrams, not only can he do away with disconnected graphs altogether, he can also achieve yet another reduction of the description to its essential elements. To this end Dyson defines a "self-energy part" of a Feynman diagram: A self-energy part is connected to the rest of the diagram by, roughly speaking, a single line. By deleting a self-energy part from a Feynman diagram G, one obtains a "reduced graph" G_0; and to any given Feynman diagram G_0 there is a class of Feynman diagrams related to G_0 by the addition of self-energy parts (see Fig. 6.3).

To calculate matrix elements, one should take into account all these Feynman diagrams. Dyson can show, however, that it is sufficient to take into account the "totally reduced graph" (p. 498) if the translation of the elements of the graph into mathematical expressions is modified. Dyson explains the procedure as follows:

> A "self-energy part" of a graph G is defined as follows; it is a set of one or more vertices not including x_0, together with the lines joining them, which is connected with the remainder of G (or with the edge of the diagram) only by two electron lines or by one or two photon lines. For definiteness it may be supposed that G has a self-energy part F, which is connected with its surroundings only by one electron line entering F at x_1, and another leaving F at x_2 [see Fig. 6.3]; the case of photon lines can be treated in an entirely analogous way. The points x_1 and x_2 may or may not be identical. From G a "reduced graph" G_0 can be obtained by omitting F completely and joining the incoming line at x_1 with the outgoing line at

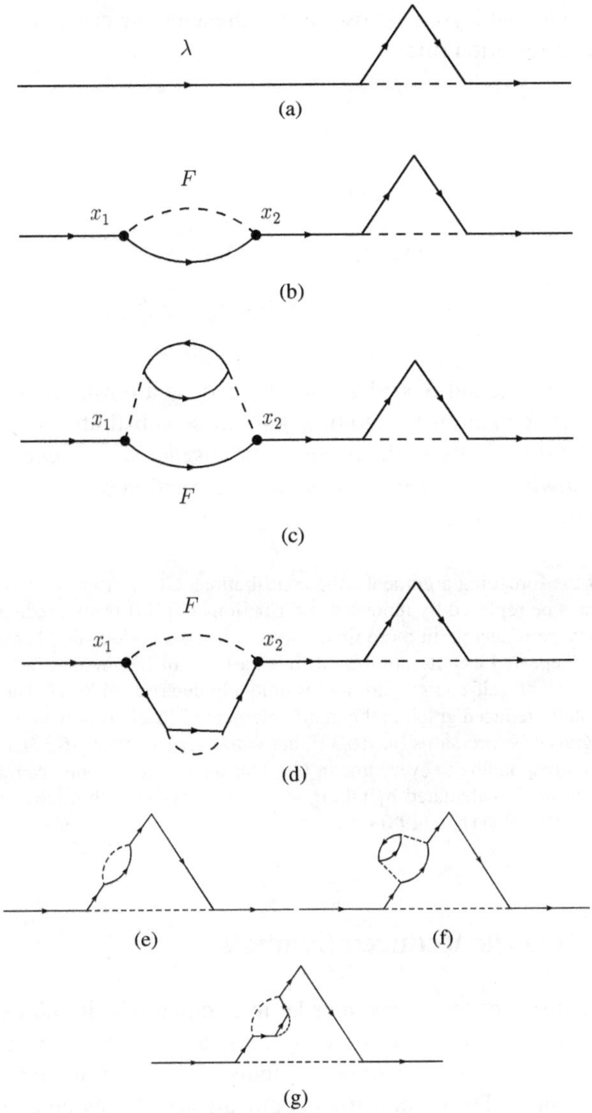

Fig. 6.3 Some of the infinitely many types of graphs that do not have to be explicitly taken into account if the "totally reduced" graph G_0 (**a**) is translated into mathematical expressions according to the modified translation rules of (6.37), (6.38), (6.39), and (6.40). Graphs (**b**)–(**d**) show the insertion of self-energy parts into the incoming line of G_0, while (**e**)–(**g**) show the insertion of the same self-energy parts into one of the internal lines of G_0; graphs by A. W.

x_2 to form a single electron line in G_0, the newly formed line being denoted by λ [see Fig. 6.3]. Given G_0 and λ, there is conversely a well determined set Γ of graphs G which are associated with G_0 and λ in this way; G_0 itself is considered also to belong to Γ. It will now be shown that the sum $C(\Gamma)$ of the contributions $C(G)$ to the matrix element of [(6.21)] from all the graphs G of Γ reduces to a single term $C'(G_0)$. (RadTh, p. 496)

For the example that Dyson chooses as a representative case, the necessary modifications of the translation rules are

$$\bar{\psi}(x_3) \rightarrow R_1 \bar{\psi}(x_3), \tag{6.37}$$

$$\psi(x_3) \rightarrow R_1^* \psi(x_3), \tag{6.38}$$

$$S_F(x_3 - x_4) \rightarrow S_F'(x_3 - x_4), \tag{6.39}$$

$$D_F(x_3 - x_4) \rightarrow D_F'(x_3 - x_4), \tag{6.40}$$

where R_1 is a constant and S' and D' are defined by a power series in terms of S and D (see RadTh, equations 62 and 67). With these substitutions the contributions of all graphs that differ only in the presence or absence of self-energy parts can be obtained by following the procedure described in Section 6.2.2 to find the relevant terms of the matrix element:

> As a result of the foregoing arguments, the contributions $C(G)$ of graphs with self-energy parts can always be replaced by modified contributions $C'(G_0)$ from a reduced graph G_0. A given G may be reducible in more than one way to give various G_0, but if the process of reduction is repeated a finite number of times a G_0 will be obtained which is "totally reduced," contains no self-energy part, and is uniquely determined by G. The contribution $C'(G_0)$ of a totally reduced graph to the matrix element of [6.21] is now to be calculated as a sum of integrals of expressions like [6.32], but with a replacement [(6.37), (6.38), (6.39), (6.40)] made corresponding to every line in G_0. This having been done, the matrix element of [(6.21)] is correctly calculated by taking into consideration each totally reduced graph once and once only. (RadTh, p. 498)

6.5 Interpreting the Reduced Symbols

Dyson introduces self-energy parts in order to eliminate the infinities related to the self-energy of the electron. Up to now Dyson's discussion has centred mainly on the syntactical level of the two symbol systems: "Feynman diagrams" and "mathematical expressions". Dyson constructs equivalence classes containing infinitely many graphs: graphs that differ from each other only by disconnected parts and self-energy parts belong to the same class. Dyson shows that it is sufficient to take into account only one representative element of a class, the "totally reduced graph", instead of infinitely many of them. The mathematical expression that corresponds to the contribution of the whole class is obtained by applying the modified translation rules to the representative element. He then goes on to interpret the self-energy parts of the graph as physical processes and the corresponding mathematical expressions as a description of the quantitative contribution of those processes to the matrix elements or, finally, transition probabilities.

6.5.1 Mass Renormalization

Dyson first considers a particular type of self-energy part, namely those self-energy parts for which the points x_1 and x_2 coincide. Dyson explicitly allows for this possibility in the passage, quoted above, in which he introduces self-energy parts: "The points x_1 and x_2 may or may not be identical." If the points coincide, the self-energy part consists of just this one point (see Fig. 6.4). This type of self-energy part corresponds to the mathematical expression

$$H^{\mathrm{S}}(x) = -\delta m\, c^2 \bar{\psi}(x)\psi(x) \tag{6.41}$$

(see (6.13)).

Dyson introduces these terms to let the state vector Ω describe a particle with the *observable* mass and, as a consequence, *finite* self-energy (see end of Section 6.2.1). Otherwise, a consistent definition of an appropriate state vector would have been impossible. Together with the other terms arising from H^{i} (superscript lower-case "i"), these terms have, as a net result, an adjustment, or *renormalization*, of the mass. One can, therefore, omit the self-energy parts corresponding to all these terms and compensate the omission simply by replacing the bare mass parameter m with the observable mass:

> The elimination of the graphs with self-energy parts is a most important simplification of the theory. For according to [(6.13)], H^{I} contains the subtracted part H^{S}, which will give rise to many additional terms in the expansion of [(6.21)]. But if any such term is taken, say, containing the factor $H^{\mathrm{S}}(x_i)$ in the integrand, every graph corresponding to that term will contain the point x_i joined to the rest of the graph only by two electron lines, and this point by itself constitutes a self-energy part of the graph. Therefore, all terms involving H^{S} are to be omitted from [(6.21)] in the calculation of matrix elements. The intuitive argument for omitting these terms is that they were only introduced in order to cancel out higher order self-energy terms arising from H^{i}, which are also to be omitted; the analysis of the foregoing paragraphs is a more precise form of this argument. In physical language, the argument can be stated still more simply; since δm is an unobservable quantity, it cannot appear in the final description of observable phenomena. (RadTh, p. 498)

6.5.2 Introducing a New Type of Vacuum Polarization

Dyson continues with his interpretation of the self-energy parts in section VIII of the article (RadTh). Apart from the effect taken into account by mass renormalization, other adjustments expressed by (6.38), (6.39), (6.40) are made to compensate for

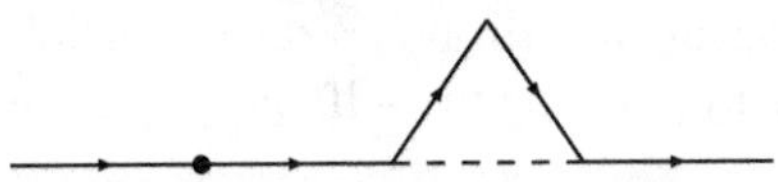

Fig. 6.4 Graph (by A. W.) with a self-energy part, with x_1 and x_2 identical. The corresponding reduced graph is the graph of Fig. 6.3a

the omission of the self-energy parts. According to Dyson, these adjustments take into consideration the observable effects of the vacuum polarization that the omitted graphs would represent:

> The question now arises: What is the physical meaning of the new functions D'_F and S'_F, and of the constant R_1? In general terms, the answer is clear. The physical processes represented by the self-energy parts of the graph have been pushed out of the calculations, but these processes do not consist entirely of unobservable interactions of single particles with their self-fields, and so cannot entirely be written off as "self-energy processes". In addition, these processes include the phenomenon of vacuum polarization, i. e., the modification of the field surrounding a charged particle by the charges which the particle induces in the vacuum. Therefore, the appearance of D'_F, S'_F, and R_1 in the calculations may be regarded as an explicit representation of the vacuum polarization phenomena which were implicitly contained in the processes now ignored. (RadTh, p. 498)

Dyson distinguishes between several types of vacuum polarization. First of all he differentiates between *external* and *internal* vacuum polarization. For the moment he is only concerned with the internal type, that is, the one produced without an external field (p. 498).

Dyson first elaborates on the interpretation of D'_{F}, which he borrows from the original context of the D_{F} function. The D_{F} function, Feynman's δ_+, is a modification and reinterpretation of the mathematical expressions that describe the interaction potential in classical electrodynamics:

> To form a concrete picture of the function D'_F, it may be observed that the function $D_F(y-z)$ represents in classical electrodynamics the retarded potential of a point charge at y acting upon a point charge at z, together with the retarded potential of the charge at z acting on the charge at y. Therefore, D_F may be spoken of loosely as "the electromagnetic interaction between two point charges". In this semiclassical picture, D'_F is then the electromagnetic interaction between two point charges, including the effects of the charge-distribution which each charge induces in the vacuum. (RadTh, pp. 498–499)

The example Dyson gives is Møller scattering,[18] that is, the scattering of two electrons, which is, incidentally, the process represented by the "first" Feynman diagram (see Fig. 5.8). A slightly more abstract form of the diagram is shown in Fig. 6.5. The change from D_{F} to D'_{F} takes into account that the simple exchange of one photon is not the only process involved. The exchanged photon may, for instance, produce a virtual electron–positron pair (see Fig. 6.5b). By such processes, represented by self-energy parts, the strength of interaction between two electrons is modified. Replacing D_{F} with D'_{F} in the calculation of the transition probability amplitude takes into consideration this modification. The replacement compensates for the omission of graphical elements that describe vacuum polarization processes, which one usually should include in a description of Møller scattering. The graphs that would include the vacuum polarization processes are obtained from the reduced graph shown in Fig. 6.5a by inserting the self-energy parts into the photon line (see, for example, Fig. 6.5b).

---

[18] See, e. g., Møller (1931, 1932); Roqué (1992).

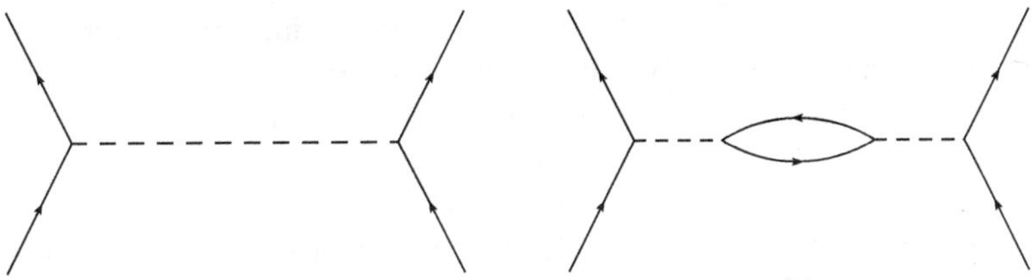

(a) (b)

Fig. 6.5 The translation of graph (**a**) into the mathematical expressions for the probability amplitude for Møller scattering using D_F gives the amplitude when vacuum polarization is ignored. If graph (**a**) is taken to be the "totally reduced graph", representative of the class of graphs obtained from it by the insertion of photon self-energy parts, it is translated using D'_F and takes into account the effects of vacuum polarization. Graph (**b**) is one of the infinitely many members of the mentioned class; graphs by A. W.

The example illustrates that (internal) vacuum polarization effects are entirely taken into account by replacing D_F with D'_F. However, this and mass renormalization were not the only adjustments necessary to compensate for the omission of self-energy parts. For instance, we saw that S_F should be replaced with S'_F when omitting the self-energy parts. Which physical process does this replacement take into account? In order to establish an interpretation for S'_F, Dyson has to extend the concept of vacuum polarization:

> The complete phenomenon of vacuum polarization as hitherto understood is included in the above picture of the function D'_F. There is nothing left for S'_F to represent. Thus, one of the important conclusions of the present theory is that there is a second phenomenon occurring in nature, included in the term vacuum polarization as used in this paper, but additional to vacuum polarization in the usual sense of the word. (RadTh, p. 499)

The symbol system "Feynman diagrams" shows how similar Møller scattering (Fig. 6.5) is to another process, namely *Compton scattering* (see Fig. 6.6). In Møller scattering, the function D_F can be readily interpreted as describing the interaction between two electrons, drawing on the analogy to the function representing the retarded potential in classical electrodynamics. No classical potential exists that would suggest a similar interpretation for S_F. But the similarity in structure of the Feynman diagrams representing Møller and Compton scattering provides Dyson with the necessary analogy. As D_F in Møller scattering, so S_F would represent the

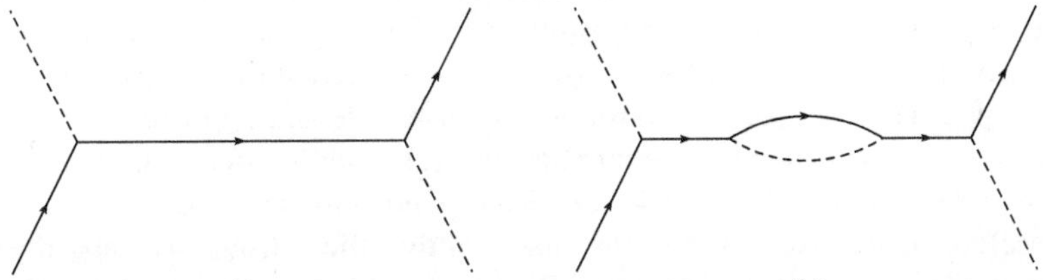

(a) (b)

Fig. 6.6 Graphs representing Compton scattering: reduced graph (**a**) and graph with a self-energy part (**b**); graphs by A. W.

interaction energy in the case of Compton scattering. Since S_F corresponds to an electron line in the graph, S'_F takes into account the effect of the alterations of the interaction brought about by electron self-energy parts. Because of the similarity between the representation of Møller scattering and Compton scattering, Dyson conceives of the effect as a type of vacuum polarization, too. Dyson's representation and interpretation of Compton scattering are new and have little to do with finding a more convenient method of calculation. Rather, the new representation of what has allegedly been known for some time leads him to recognize that he is representing a new physical content:

> The scattering of one electron by another may be represented as caused by a potential energy (the Møller interaction) acting between them. If one electron is at y and the other at z, then, as explained above, the effect of vacuum polarization of the usual kind is to replace a factor D_F in this potential energy by D'_F. Now consider an analogous, but unorthodox, representation of the Compton effect, or the scattering of an electron by a photon. If the electron is at y and the photon at z, the scattering may be again represented by a potential energy, containing now the operator $S_F(y-z)$ as a factor; the potential is an exchange potential, because after the interaction the electron must be considered to be at z and the photon at y, but this does not detract from its usefulness. By analogy with the 4-vector charge-current density j_μ, which interacts with the potential D_F, a spinor Compton-effect density u_α may be defined [...]
>
> These spinors are not directly observable quantities, but the Compton effect can be adequately described as an exchange potential, of magnitude proportional to $S_F(y-z)$, acting between the Compton-effect density at any point y and the adjoint density at z. The second vacuum polarization phenomenon is described by a change in the form of this potential from S_F to S'_F. Therefore, the phenomenon may be pictured in physical terms as the inducing, by a given element of Compton-effect density at a given point, of additional Compton-effect density in the vacuum around it. (RadTh, p. 499)

Through using Feynman diagrams, Dyson understands Compton scattering, that is, the scattering of a photon off an electron, to be brought about by the absorption and emission of the photon by the electron. The strength of interaction is modified by, for instance, the possibility that in the intermediate state the electron may emit and re-absorb a virtual electron. Replacing S_F with S'_F takes into account this modification quantitatively.

6.5.3 Charge Renormalization

What remains to be interpreted—after mass renormalization and replacing D_F and S_F with D'_F and S'_F—is the last adjustment: introducing the constants R_1, R_2 and R_3. R_1 occurs when the modified translation rules, needed to take into account the self-energy effects (see (6.37) and (6.38)), are used; R_2 and R_3 occur when D'_F and S'_F, which are also needed to take into account self-energy effects, are used instead of D_F and S_F, (see (6.39) and (6.40) and RadTh, equations 62 and 67).

These constants take into account that omitting the self-energy parts does not only alter the shape of the functions D_F and S_F but also changes their absolute value. The functions are scaled by the factors R_1, R_2 and R_3. All three parameters are infinite, which means that they cannot be interpreted as a physical quantity.

However, similar to the way in which infinite self-energy effects are cancelled by introducing an infinite mass parameter δm, the charge parameter e, occurring in the Hamiltonian that describes the interaction (see (6.9) and (6.23)), is recognized as the charge of an electron in the hypothetical case of no vacuum polarization (compare the mass of the "bare" electron). A combination of the parameter e and the infinite parameters R equals the finite observed charge of the electron. The trading in of the "bare" charge for the observed charge is called *charge renormalization*:

> In both sorts of internal vacuum polarization, the functions D_F and S_F, in addition to being altered in shape, become multiplied by numerical (and actually divergent) factors R_3 and R_2; also the matrix elements of [6.21] become multiplied by numerical factors such as $R_1 R_1^*$. However, it is believed (this has been verified only for second-order terms) that all nth-order matrix elements of [6.21] will involve these factors only in the form of a multiplier
>
> $$\left(eR_2R_3^{\frac{1}{2}}\right)^n; \tag{6.42}$$
>
> [...] Here e is defined as the constant occurring in the fundamental interaction [6.9] by virtue of [6.23]. Now the only possible experimental determination of e is by means of measurements of the effects described by various matrix elements of [6.21], and so the directly measured quantity is not e but $eR_2R_3^{\frac{1}{2}}$. Therefore, in practice the letter e is used to denote this measured quantity, and the multipliers R no longer appear explicitly in the matrix elements of [6.21]; the change in meaning of the letter e is called "charge renormalization", and is essential if e is to be identified with the observed electronic charge. (RadTh, p. 499)

Dyson's outlined procedure does not, of course, evaluate the functions D'_{F} and S'_{F}, which will be necessary if the matrix elements of the operator H_{F} given in (6.21) are to be calculated. However, by defining the equivalence classes of graphs that differ only in the insertions of self-energy parts, Dyson has reduced the problem of evaluating a plethora of mathematical expressions, one from each graph in the equivalence class, to the evaluation of the two single expressions D'_{F} and S'_{F} that occur in the mathematical expression corresponding to the one representative member, the totally reduced graph. Moreover, the problem of evaluating D'_{F} and S'_{F} is of the same type as the superordinate problem of determining the matrix elements of H_{F}:

> The determination of the [D'_F and S'_F] is a problem of the same kind as the original problem of the calculation of matrix elements of [6.21], and the various terms in the [operators of which D'_F and S'_F are the vacuum expectation values] must again be split up, represented by graphs, and analyzed in detail. However, since D'_F and S'_F are universal functions, this further analysis has only to be carried out once to be applicable to all problems. (RadTh, p. 500)

6.6 The Removal of the Infinities

In his first paper, 'The Radiation Theories of Tomonaga, Schwinger, and Feynman' (RadTh), Dyson was mainly concerned with the proof of equivalence of Schwinger's and Feynman's theories. Having thus clarified Feynman's method and established its validity, Dyson turns to a more comprehensive treatment of the problematic infinities of QED. It is in this second paper, 'The *S* Matrix in Quantum Electrodynamics' (SM), that Dyson clearly demonstrates the power of Feynman diagrams and their crucial role in completely removing the uninterpretable infinities.

In SM, Dyson extends his systematization of Feynman's methods to include the treatment of problems with several particles in the initial and final states, instead of only one as in RadTh. He also now explicitly presents Feynman's theory as an *S matrix* theory in the tradition of Werner Heisenberg,[19] which makes it suitable for treating scattering problems but not problems involving bound states. When several particles are involved, the theoretical quantities which are of interest are the matrix elements of the operator $U(\infty)$, whereas in RadTh, where Dyson mainly considered one-particle states, these were the matrix elements of the operator H_F (SM, section II).

First, in a summary of the discussion of RadTh Dyson states that matrix elements can be systematically calculated using a one-to-one correspondence between the elements of a graph and mathematical expressions such as D_F and S_F. Translating graphs into mathematical expressions is the essential part of the "Feynman rules for the evaluation of $U(\infty)$" (SM, p. 1738).

Unlike in the first paper, in SM Dyson proceeds more systematically to the problem of eliminating the infinities. While in RadTh removing the infinities and introducing the new symbol system of Feynman diagrams were, to a large extent, interlinked, Dyson now clearly separates the two tasks:

> The above rules for the calculation of $U(\infty)$ describe the state of affairs before any attempt has been made to identify and remove the various divergent parts of the expressions. In particular, contributions are included from all graphs *G*, even those which yield nothing but self-energy effects. For this reason, the rules here formulated are superficially different from those given for the one-electron problem in Section IX of [RadTh], which described the state of affairs after many divergencies had been removed. Needless to say, the rules are not complete until instructions have been supplied for the removal of all infinite quantities from the theory; in Sections V–VII of this paper [SM] it will be shown how the formal structure of the *S* matrix makes such a complete removal of infinities appear attainable. (SM, p. 1739)

In SM Dyson also mainly uses momentum variables rather than space-time variables for the first time, an alternative that Feynman had introduced in ThPos and STQED. However, Dyson's choice of variables is not my concern here. My aim is to describe and analyze the function of Feynman diagrams in Dyson's method, to which it makes little difference whether Dyson uses momentum or space-time variables. More important is the fact that Dyson introduces a new equivalence relation between graphs, which enables him to make the algorithm for evaluating matrix elements still more effective and allows for fewer redundancies in the diagrams' articulation of the phenomena.

---

[19] See, e. g., Heisenberg (1943) (cited in RadTh, fn. 5, p. 489).

6.6.1 The Further Removal of Redundant Symbols

Vertex parts In RadTh, Dyson defines a class of graphs that differ only in the presence or absence of self-energy parts (see Section 6.4.2). To calculate the matrix elements, it then suffices to consider the totally reduced graph, that is, the graph in the class that does not contain any self-energy parts. Now (in SM) Dyson also incorporates into the equivalence class of a graph all the graphs that differ from it only by what he calls the *vertex parts*:

> By a "vertex part" of any graph will be meant a connected part of the graph, consisting of vertices and internal lines only, which touches precisely two electron lines and one photon line belonging to the remainder of the graph. The central triangle of Fig. 2 [Fig. 6.7b] is an example of such a part. In other words, a vertex part of a graph is a part which can be substituted for the single vertex of Fig. 1 [Fig. 6.7a] and give a physically meaningful result. (SM, p. 1741)

If a graph G' is obtained from a graph G by inserting a vertex part V into one of G's vertices, the mathematical expression M' corresponding to G' is obtained from the mathematical expression M corresponding to G by replacing the γ_μ associated with the vertex in G with $\Lambda_\mu(V, t^1, t^2)$.

Dyson recalls the similar results implicit in RadTh concerning self-energy parts: if a graph G' is obtained from a graph G by inserting a self-energy part W into one of its internal electron lines, the mathematical expression M' corresponding to G', is obtained from the mathematical expression M corresponding to G by replacing the S_F associated with the internal electron line in G with $S_F(t^1)\Sigma(W, t^1)S_F(t^1)$. If the electron line is external, a factor $\bar{\psi}(t^1)$ has to be replaced with a factor $\bar{\psi}(t^1)\Sigma(W, t^1)S_F(t^1)$. If the self-energy part is the special case consisting of just one vertex (see Fig. 6.4), $\Sigma(W, t^1)$ is just $-2\pi i(\delta mc/\hbar) = -2\pi i\delta\kappa_0$. For self-energy parts W' inserted into photon lines, the replacements are $D_F(t^3)\Pi(W', t^3)D_F(t^3)$ for $D_F(t^3)$; $A_\mu(t^3)\Pi(W', t^3)D_F(t^3)$ for $A_\mu(t^3)$.

An important feature of these replacements is that they only depend on the type of vertex part V or self-energy part W or W' inserted and on the energy-momentum four-vectors t^1 or t^3 associated with the concerned lines. The replacements are

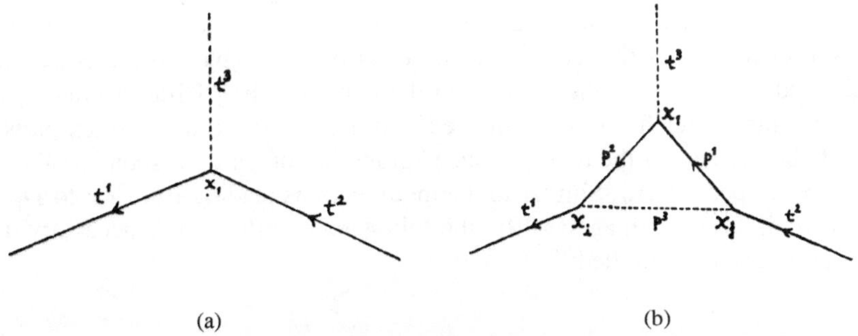

Fig. 6.7 Introducing the vertex part of a graph (Figures 1 and 2 of SM). (**a**) The vertex of a graph. (**b**) The vertex part replacing the simple vertex of Fig. 6.7a

universal in the sense that they are the same, irrespective of the graph into which they are inserted.

Unlike self-energy parts, the treatment of disconnected parts (see Section 6.4.1) cannot be adopted from RadTh without modifications being made, since, in the more general case of several particles considered now, not all the disconnected parts can be regarded as irrelevant:

> It was shown in Section VII of [RadTh] that, for the one-electron processes there considered, only connected graphs needed to be taken into account. In constructing the S matrix in general, this is no longer the case; disconnected graphs give matrix elements of $U(\infty)$ representing two or more collision processes occurring simultaneously among separate groups of particles, and such processes have physical reality. It is only permissable to omit a disconnected graph when one of its connected components is entirely lacking in external lines; such a component without external lines will give rise only to a constant multiplicative phase factor in every matrix element of $U(\infty)$ and is therefore devoid of physical significance. (SM, p. 1742)

Furthermore, in contrast to RadTh, Dyson now does not differentiate between an external electromagnetic potential $A^{\mathrm{e}}_{\mu}(x_0)$ and the electromagnetic field produced by the particles:

> In evaluating $U(\infty)$ we shall not make any distinction between the external and radiative parts of the electromagnetic field; this is physically reasonable since it is to some extent a matter of convention how much of the field in a given situation is to be regarded as "external". (SM, p. 1738)

With these provisos and the introduction of momentum variables and vertex parts, Dyson is ready to extend his algorithm for finding relevant matrix elements to the general case:

> In Section VII of [RadTh], it was shown how self-energy parts could be systematically eliminated from all graphs, and their effects described by suitably modifying the functions D_F and S_F. The analysis was carried out in configuration space, and was confined to the one-electron problem. We are now in a position to extend this method to the whole S matrix formalism, working in momentum space, and furthermore to eliminate not only self-energy parts but also the "vertex parts" defined in the last section. (SM, pp. 1742/1743)

As with the self-energy parts already defined in RadTh, Dyson shows that only the graph without vertex parts has to be considered *if* the correspondence rules between the graphical elements and the mathematical expressions are modified. With self-energy parts this concerned the lines of the graph and the corresponding D_{F} and S_{F}; now it is the vertex and the corresponding Dirac operator γ_{μ}. If the graphs that differ from the totally reduced graph only in their vertex parts are omitted, the vertices of the totally reduced graph no longer correspond to a simple γ_{μ} but to an operator Λ_{μ}. Similar to the modifications leading from D_{F} to D'_{F} and from S_{F} to S'_{F} (see (6.39) and (6.40)), the following modification is necessary if the vertex parts are to be omitted[20]:

[20] SM, equation 38.

$$\gamma_\mu \to \Gamma_\mu(t^1, t^2) = \gamma_\mu + \Lambda_\mu(t^1, t^2), \tag{6.43}$$

where $\Lambda_\mu(t^1, t^2)$ is the sum of all $\Lambda_\mu(V, t^1, t^2)$s associated with the insertion of all possible vertex parts V:

$$\Lambda_\mu(t^1, t^2) = \sum_V \Lambda_\mu(V, t^1, t^2). \tag{6.44}$$

The generalized replacement rule for the internal electron line is[21]:

$$S_{\mathrm{F}}(p^i) \to S'_{\mathrm{F}} = S_{\mathrm{F}}(p^i) + S_{\mathrm{F}}(p^i)\Sigma(p^i)S_{\mathrm{F}}(p^i), \tag{6.45}$$

where $\Sigma(p^i) = \sum_W \Sigma(W, p^i)$ and p^i the energy-momentum four-vector associated with the line. For the internal photon line, the replacement rule is[22]:

$$D_{\mathrm{F}}(p^i) \to D'_{\mathrm{F}} = D_{\mathrm{F}}(p^i) + D_{\mathrm{F}}(p^i)\Pi(p^i)D_{\mathrm{F}}(p^i), \tag{6.46}$$

where $\Pi(p^i) = \sum_{W'} \Pi(W', p^i)$. For the external lines (ingoing or outgoing electron or photon), the replacement rules are[23]:

$$\psi(k^i) \to \psi'(k^i) = S_F(k^i)\Sigma(k^i)\psi(k^i) + \psi(k^i), \tag{6.47}$$
$$\bar{\psi}(k^i) \to \bar{\psi}'(k^i) = \bar{\psi}(k^i)\Sigma(k^i)S_F(k^i) + \bar{\psi}(k^i), \tag{6.48}$$
$$A_\mu(k^i) \to A'_\mu(k^i) = A_\mu(k^i)\Pi(k^i)D_F(k^i) + A_\mu(k^i), \tag{6.49}$$

where k^i is like t^i and p^i, the energy-momentum four-vector associated with the lines.

In RadTh, Dyson gives a physical interpretation of S'_{F} and D'_{F}, which compensates for the omission of self-energy parts (see Section 6.5). Now he also gives a physical interpretation for Γ_μ, which compensates for the omission of the vertex parts:

> The functions D'_F and S'_F of [(6.45)] and [(6.46)] are the Fourier transforms of the corresponding functions in [RadTh]. The interpretation of these functions in Section VIII of [RadTh] can be extended in an obvious way to include the operator Γ_μ. Since $\bar{\psi}\gamma_\mu\psi$ is the charge-current 4 vector of an electron without radiative corrections, $\bar{\psi}\Gamma_\mu\psi$ may be interpreted as an "effective current" carried by an electron, including the effects of exchange interactions between the electron and the electron–positron field around it. (SM, p. 1743)

Odd parts Dyson has still not eliminated all the redundancies in the Feynman diagrams' articulation of quantum electrodynamic phenomena. In the diagrams'

---

[21] SM, equation 35.

[22] SM, equation 36.

[23] SM, equation 37.

symbol system, a theorem proved by Wendell Furry[24] is reflected in the omission of graphs that have, what Dyson calls, "odd parts":

> An additional reduction in the number of graphs effectively contributing to $U(\infty)$ is obtained from a theorem of Furry [Furry 1937]. The theorem was shown by Feynman to be an elegant consequence of his theory.
>
> [...]
>
> By an "odd part" of a graph is meant any part, consisting only of vertices and internal lines, which touches no electron lines, and only an odd number of photon lines, belonging to the rest of the graph. The simplest type of odd parts which can occur is a single odd closed loop. Conversely, it is easy to see that every odd part must include within itself at least one odd closed loop. Therefore, Furry's theorem allows all graphs with odd parts to be omitted from consideration in calculating $U(\infty)$. (SM, pp. 1743/1744)

The simplest type of even and odd parts are shown in Fig. 6.8.

Feynman had explained the theorem only in passing as being a consequence of the fact that an electron propagating backwards has the same effects as a positron propagating ordinarily:

> Loops with an odd number of potential interactors give zero. Physically this is because for each loop the electron can go around one way or in the opposite direction and we must add these amplitudes. But reversing the motion of an electron makes it behave like a positive charge thus changing the sign of each potential interaction, so that the sum is zero if the number of interactions is odd. This theorem is due to [Furry 1937]. (ThPos, fn. 9)

6.6.2 *Isolating the Infinities*

Dyson has not yet removed all the irrelevant graphs as far as the isolation and eventual removal of the uninterpretable infinities are concerned. He shows that only four types of graph need to be investigated for their divergence properties if one is to be able to eliminate the divergences from all graphs.

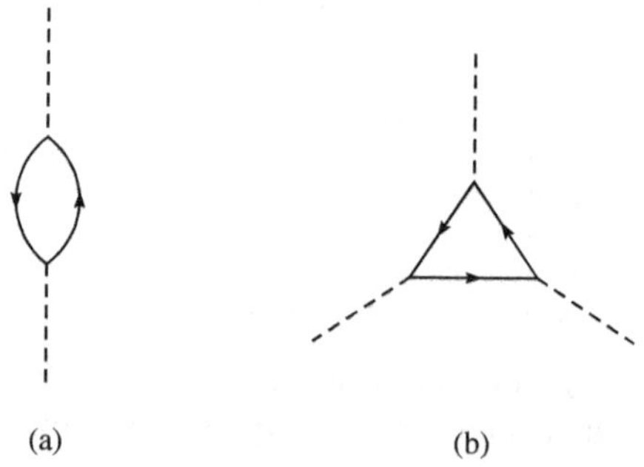

Fig. 6.8 Closed loops with an even or odd number of interactions with a photon of an electromagnetic potential (internal or external); graphs by A. W. (**a**) Even closed loop: two interactions. (**b**) Odd closed loop: three interactions

[24] Furry (1937).

To this end, Dyson introduces for the first time the notion of a *primitive* divergent graph G and matrix element M:

> A divergent M is called "primitive" if, whenever one of the momentum 4 vectors in its integrand is held fixed, the integration over the remaining variables is convergent. Correspondingly, a primitive divergent graph is a connected graph G giving rise to divergent M, but such that, if any internal line is removed and replaced by two external lines, the modified G gives convergent M. To analyze the divergences of the theory, it is sufficient to enumerate the primitive divergent M and G and to examine their properties. (SM, p. 1745)

Figure 6.9 shows an example of a primitive divergent graph and the convergent graph that results from replacing the internal electron line with an incoming line and an outgoing external line. The convergent graph is identical to the first Feynman diagram representing Møller scattering (see Figs. 5.8 and 6.5a).

Dyson finds that the divergence properties of M are reflected in a fraction of the momentum variables contained within it. Accordingly, a convergence condition for M can be stated in terms of the degree of powers in which the momentum variables occur in the numerator and denominator. For any primitive divergent graph G with n vertices Dyson finds the following condition for convergence:

$$K = 2F - F_e - 4[F - n + 1] \geq 1, \tag{6.50}$$

where $2F$ is the degree of the denominator and F_e the degree of the numerator.

The degrees of the relevant denominator and numerator are reflected in the structure of graph G. This, and the irrelevancy of graphs with odd parts, will allow Dyson to sort out the relevant graphs:

> Let E_e and E_p be the numbers of external electron and photon lines in G, and let n_s be the number of vertices without photon lines incident. It follows from the structure of G that
>
> $$2F = 3n - n_s - E_e - E_p, \tag{6.51}$$
>
> $$F_e = n - \tfrac{1}{2}E_e, \tag{6.52}$$
>
> and so the convergence condition [(6.50)][25] is

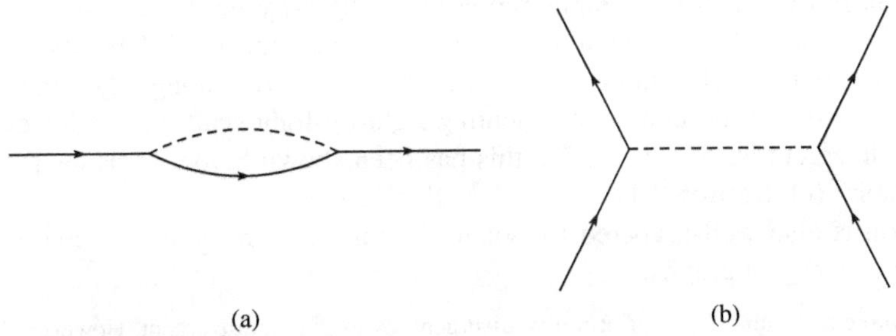

Fig. 6.9 A primitive divergent graph (a) and a convergent graph (b), which results from replacing the internal electron line with two external lines; graphs by A. W.

[25] The original text refers the reader to the wrong equation, that is, equation 52 instead of equation 54 in SM.

$$K = \frac{3}{2}E_e + E_p + n_s - 4 \geq 1. \tag{6.53}$$

> This gives the vital information that the only possible primitive divergent graphs are those with $E_e = 2$, $E_p = 0, 1$, and with $E_e = 0$, $E_p = 1, 2, 3, 4$. Further, the cases $E_e = 0$, $E_p = 1, 3$, do not arise, since these give graphs with odd parts which were shown to be harmless in Section IV. (SM, p. 1746)

The graphs that Dyson thus identifies as problematic represent physical processes that are well-known to physicists, for instance, light-by-light scattering (cf. Section 2.1):

> The possible primitive divergent graphs that have been found are all of a kind familiar to physicists. The case $E_e = 2$, $E_p = 0$ describes self-energy effects of a single electron; $E_e = 0$, $E_p = 2$ self-energy effects of a single photon; $E_e = 2$, $E_p = 1$ the scattering of a single electron in an electromagnetic field; and $E_e = 0$, $E_p = 4$ the "scattering of light by light" or the mutual scattering of two photons. (SM, p. 1746)

The convergence condition (6.53) also shows that the divergences are of three weak types, which assures Dyson that he is on the point of isolating the problematic divergences. And if a way can be found to eliminate the latter, then all the possibly occurring infinities in QED can also be removed:

> Further, [(6.53)] shows that the divergence will never be more than logarithmic in the third and fourth cases, more than linear in the first, or more than quadratic in the second. Thus it appears that, however far quantum electrodynamics is developed in the discussion of many-particle interactions and higher order phenomena, no essentially new kinds of divergence will be encountered. This gives strong support to the view that "subtraction physics", of the kind used by Schwinger and Feynman, will be enough to make quantum electrodynamics into a consistent theory. (SM, p. 1746)

The next step in isolating the problematic infinities is to rewrite the divergent quantities as a power series using momentum variables. Then, according to the degree of divergence, the infinities will be contained in the logarithmic, the constant, or the linear term of the power series. The rest of the series will be a finite quantity, and an unambiguous separation of the infinities from the finite parts of the problematic expressions will be achieved.

The primitive divergent graphs representing light-by-light scattering are easily done, since, on closer inspection, they are not divergent at all. The above procedure to sort out the relevant graphs was thus not restrictive enough. Dyson provides his own proof that the graphs representing light-by-light scattering to lowest order are not divergent but mentions that this has been shown before by Hans Euler and B. Kockel[26] (cf. Section 2.1).

Dyson is also well-prepared to isolate the infinities in the other three types of primitive divergent graphs:

> The three remaining types of primitive divergent M are, in fact divergent. However, these are just the expressions which have been studied in Sections III and IV and shown to be completely described by the operators Λ_μ, Σ, and Π.
>
> [...]

[26] SM, fn. 9; Euler and Kockel (1935); Euler (1936).

> [If] some means can be found for isolating and removing the divergent parts from Λ_μ, Σ, and Π, the "irreducible graphs" defined in Section IV will not introduce any fresh divergences into the theory, and the rules of Section IV will lead to a divergence-free S matrix. (SM, p. 1747)

Even after having separated the infinite from the finite parts in Λ_μ, Σ and Π, the infinities cannot yet be satisfactorily removed. Omitting the infinities at this stage would leave as finite only a very crude approximation of what is believed to be the correct interpretable expression for Γ_μ, S'_F and D'_F, determined by Λ_μ, Σ and Π (see (6.43), (6.45) and (6.46)). To obtain increasingly better approximations, the separation has to be conducted repeatedly, since self-energy or vertex parts can be reintroduced into the lines and vertices of, for instance, the electron self-energy parts, omitted and compensated by S'_F, and in the lines and vertices of the newly introduced parts again, and so forth. As the appropriate elements have been parsed, the same steps can be repeated again and again and no new difficulties will arise:

> The task remaining is to complete the formulas [...], which show how the infinite parts can be separated from the operators Γ_μ, S'_F and D'_F, and to include the corrections introduced into these operators by the radiative reactions which they themselves describe. In other words, we have to include radiative corrections to radiative corrections, and renormalizations of renormalizations, and so on *ad infinitum*. This task is not so formidable as it appears. (SM, p. 1749)

The infinite parts of the operators Γ_μ, S'_F and D'_F are dropped at each stage of the procedure, and the finite part will approach the exact solution more closely the more frequently the steps are repeated.[27] However, Dyson still needs to explain why this procedure will end in a correct, as well as finite, result. After all, he is discarding terms that were derived from the theory's basic equations. Such a procedure could be suspected of being arbitrary, and indeed similar methods came to be known pejoratively as "subtraction physics" (a term that also Dyson uses in one of the above quotations).[28]

Dyson justifies his method of subtracting the infinities by pointing out that dispensing with the infinities corresponds to replacing the mass and charge parameters in the basic equations of the theory with their empirical values:

> It is necessary finally to justify the dropping of the divergent terms. This will be done by showing that the "true" Γ_μ, S'_F and D'_F, which are obtained if the divergent terms are not dropped, are only numerical multiples of those obtained by dropping divergences, and that the numerical multiples can themselves be eliminated from the theory by a consistent use of the ideas of mass and charge renormalization. (SM, p. 1750)

The exact expressions for Γ_μ, S'_F and D'_F, that is, the modified operators that would compensate for *all* omitted self-energy and vertex parts, are those expressions that, in a further step of the approximation procedure, would no longer change their values. These operators are infinite. The finite operators $\Gamma_{\mu 1}(e)$, $S'_{\mathrm{F}1}(e)$ and $D'_{\mathrm{F}1}(e)$

---

27 SM, p. 1750.

28 See, e. g., Miller (1995), Section 4.7 and Schwinger (1983, p. 330). The expression seems to go back to Pauli's "Subtraktionsphysik" and "Limes-Akrobatik" (quoted in Miller 1995, p. 66).

are obtained by dropping the infinite parts at any stage of the procedure and stopping the procedure after a finite number of steps. The difference between the finite and infinite operators, however, is only one of the (infinite) multiplicative constants Z_1, Z_2 or Z_3[29]:

$$\Gamma_\mu = Z_1^{-1}\Gamma_{\mu 1}(e_1), \tag{6.54}$$

$$S'_F = Z_2 S'_{F1}(e_1), \tag{6.55}$$

$$D'_F = Z_3 D'_{F1}(e_1). \tag{6.56}$$

Here e_1 is the "true" (SM, p. 1750) electronic charge and the Zs will be determined such that their combination with the theoretical charge parameter e will give e_1, that is,[30]

$$e_1 = Z_1^{-1} Z_2 Z_3^{1/2} e. \tag{6.57}$$

Dyson proves that by choosing the appropriate Zs and the mass parameter δm, the Γ_μ, S'_{F} and D'_{F} of (6.54), (6.55), (6.56) do indeed "reproduce themselves" (p. 1750) and are thus proved to be the correct expressions. The parameter δm had already been introduced in[31] and been shown to correspond to a special type of self-energy part (see Section 6.5.1). Dyson now uses $\delta\kappa_0$, which is the same parameter in different units: $\delta\kappa_0 = \delta m\, c/\hbar$. The infinite parameters Γ_μ, S'_{F} and D'_{F} compensate for all the self-energy and vertex parts representing the myriad of inevitable virtual processes in the unobservable intermediate state of any observable process. The quantitative effect on matrix elements, or transition probabilities, of these virtual processes are all contained in the Zs:

> [If] Z_1, Z_2, Z_3, $\delta\kappa_0$ are defined by [...], it is established that [(6.54), (6.55), (6.56)] give the correct forms of the operators Γ_μ, S'_F and D'_F, including all the effects of the radiative corrections which these operators introduce into themselves and into each other. The exact Eqs. [(6.54), (6.55), (6.56)] give a much simpler separation of the infinite from the finite parts of these operators than the approximate equations [i. e. the separation in one single step in the procedure]. (SM, p. 1752)

Now Dyson can interpret the Zs that occur in the exact expressions for the infinite quantitative effect of the virtual processes on transition probabilities as a replacement of the theoretical charge parameter e by the empirical value of the charge. This is possible since, as Dyson shows, the Zs occur only in the particular combination $Z_1^{-n} Z_2^n Z_3^{\frac{1}{2}n}$, where n is the number of steps in the iterative procedure. Because of (6.57) the inverse of this combination equals the nth power of the (infinite) ratio of the theoretical charge parameter e to the empirical value e_1:

[29] SM equations 83–85.

[30] SM, equation 86.

[31] RadTh.

$$Z_1^n Z_2^{-n} Z_3^{-\frac{1}{2}n} = \left(\frac{e}{e_1}\right)^n. \tag{6.58}$$

By renormalizing the charge to its empirical value, the problematic infinities are finally eliminated in a satisfactory manner:

> If now e_1 is identified with the finite observed electronic charge, there no longer appear any divergent expressions in M. And since M is a completely general constituent of $U(\infty)$, the elimination of divergences from the S matrix is accomplished. (SM, p. 1752)

The infinities of QED had, since its first versions, threatened to make the theory uninterpretable, while how one should remove the infinities satisfactorily had been a long-standing and seemingly unsolvable problem. So the ease with which Dyson solves the problem can, therefore, come as a surprise:

> The surprising feature of the S matrix theory, as outlined in this paper, is its success in avoiding difficulties. Starting from the methods of Tomonaga, Schwinger and Feynman, and using no new ideas or techniques, one arrives at an S matrix from which the well-known divergences seem to have conspired to eliminate themselves. This automatic disappearance of divergences is an empirical fact, which must be given due weight in considering the future prospects of electrodynamics. (SM, p. 1754)

However, the ease with which Dyson finally removes the infinities is entirely comprehensible if one takes into account the importance of providing an appropriate symbol system: Dyson adopts and transforms Feynman's diagrams so that they articulate the elements and their combinations that are responsible for the infinities. Consequently, once an appropriate symbol system has been formulated, the disappearance of the divergences is, as Dyson rightly says, "automatic". The divergences' "conspiracy" is finally revealed and perceived so clearly that it appears to be "an empirical fact".

Chapter 7
The Development of a New Means of Representation: Goals and Milestones

Two of the principal aims of this study were to reconstruct the route that led Feynman, between approximately 1946 and 1948, to devise his new method of diagrams and to evaluate what was achieved. In the preceding sections I showed how Feynman developed his diagrams in a series of comprehensible steps, and that his primary consideration was to solve important theoretical problems of the theory of quantum electrodynamics (QED) as it then stood. This goal was finally achieved to a satisfactory extent by Freeman J. Dyson's systematization of Feynman's results.

7.1 The Search for Solutions to Theoretical Problems

Feynman did not take his search for alternative means of representation and interpretations of well-known equations to be an academic gimmick, nor was his main aim to speed up calculation procedures. As he wrote in a letter to his former student friend, Theodore Welton, Feynman hoped to find an alternative means of representation that would solve some of the theoretical problems of the time (see, Section 4.4):

> Of course, the hope is that a slight modification of one of the pictures will straighten out some of the present troubles.[1]

In this quote, Feynman was referring to the problematic divergences which had so far beset all attempts at finding a relativistic quantum theory for electrons and their interactions, that is, QED. Every time physicists had tried to use such a theory to calculate observable quantities to a greater precision than the leading-order contribution in the relevant quantities' perturbative expansion, the results had proved infinite and so could not, as intended, be interpreted as observable quantities.

Feynman's strategy to solve this problematic state of affairs was to find an alternative means of representation, or "picture" as he called it, that would meet two specific requirements: first, one should be able to identify the assumptions that were responsible for the contradiction between the derived results and the observed values or general theoretical principles by using the new means of representation; and, sec-

[1] *Dirac Equation a*, 12, see Fig. 4.16.

A. Wüthrich, *The Genesis of Feynman Diagrams*, Archimedes 26,
DOI 10.1007/978-90-481-9228-1_7, © Springer Science+Business Media B.V. 2010

ond, it should also be possible to modify the theoretical content so that derivations of such inconsistent results could be ruled out.

7.2 Elaboration of a Well-Known Interpretation

The central part of the pathway to the discovery of Feynman diagrams began with Feynman's representation of the quivering motion of the electron, which, according to Gregory Breit (1928) and Erwin Schrödinger (1930), is the proper physical interpretation of the Dirac equation. Feynman considered the one-dimensional case and used the two components of the wave function to describe the quivering back and forth motion of the electron in this one spatial dimension. By using his innovative representation of the electron, Feynman showed that only electrons behaving in a way that is allowed by the Dirac equation could be represented. This makes his representation a *model*, or *model system,* of the Dirac equation, in the sense, albeit roughly, of present-day philosophy of science (see, e. g., Cartwright 1983 and Giere 1996). Feynman's model is an elaboration of Breit's and Schrödinger's interpretation of the Dirac equation.

Feynman described the aforementioned electron in much the same way as one would usually use the wave function to describe the two spin degrees of freedom of the electron. In Feynman's description, the wave function is composed of the contributions of all possible paths by which the electron can reach the end point of its motion. The contributions of the individual paths depend on the number of changes in direction that the electron makes between the starting and end points of its motion. Feynman displayed graphically some representative cases of zigzag paths and, using this graphical representation, uncovered a regularity that allowed him to determine the contribution of all possible paths (see Section 4.1.3).

To make predictions about the model system of the quivering electron, one needs to know the probability of the electron travelling from the origin of the coordinate system to a given point P. In order to determine this quantity, it is important to know what kinds of possible zigzag paths exist from the origin to P , and how many reversals of direction occur in a given path. Feynman constructed the wave function of the electron by weighting each reversal by a factor $i\epsilon$ and by summing the contributions of all the paths. The absolute square of the wave function yielded the desired probability. Thus, Feynman solved the Dirac equation "by path counting" (see number 1 in Fig. 7.1).

7.3 Problematic Attempts to Generalize First Results

The next step saw Feynman attempting to integrate positrons into his model of the quivering electron. Positrons are the antiparticles of electrons, and their existence provided a solution to the interpretational problems posed by the states of negative

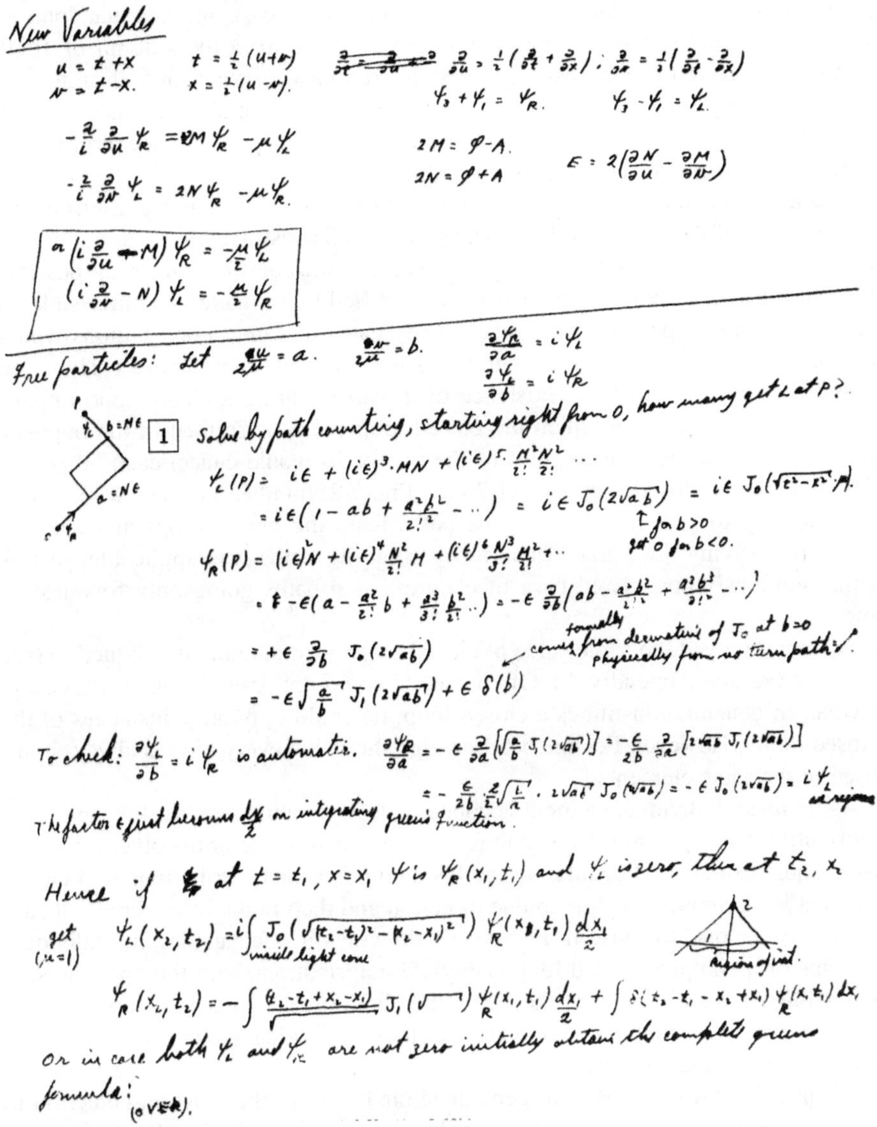
New Variables

$u = t + x \qquad t = \frac{1}{2}(u+v) \qquad \frac{\partial}{\partial u} = \frac{1}{2}\left(\frac{\partial}{\partial t} + \frac{\partial}{\partial x}\right); \quad \frac{\partial}{\partial v} = \frac{1}{2}\left(\frac{\partial}{\partial t} - \frac{\partial}{\partial x}\right)$

$v = t - x. \qquad x = \frac{1}{2}(u - v).$

$\psi_3 + \psi_1 = \psi_R. \qquad \psi_3 - \psi_1 = \psi_L.$

$-\frac{2}{i}\frac{\partial}{\partial u}\psi_R = 2M\psi_R - \mu\psi_L \qquad 2M = \varphi - A. \qquad E = 2\left(\frac{\partial N}{\partial u} - \frac{\partial M}{\partial v}\right)$

$-\frac{2}{i}\frac{\partial}{\partial v}\psi_L = 2N\psi_R - \mu\psi_R. \qquad 2N = \varphi + A$

or $\left(i\frac{\partial}{\partial u} - M\right)\psi_R = -\frac{\mu}{2}\psi_L$

$\left(i\frac{\partial}{\partial v} - N\right)\psi_L = -\frac{\mu}{2}\psi_R$

Free particles: Let $\frac{\mu u}{2} = a. \quad \frac{\mu v}{2} = b. \qquad \frac{\partial\psi_R}{\partial a} = i\psi_L \qquad \frac{\partial\psi_L}{\partial b} = i\psi_R$

1 Solve by path counting, starting right from 0, how many get L at P?

$\psi_L(P) = i\epsilon + (i\epsilon)^3 \cdot MN + (i\epsilon)^5\frac{M^2N^2}{2!\,2!}\cdots$

$= i\epsilon\left(1 - ab + \frac{a^2b^2}{2!^2} - \ldots\right) = i\epsilon J_0(2\sqrt{ab}) = i\epsilon J_0(\sqrt{t^2-x^2}\,\mu)$. for $b>0$, get 0 for $b<0$.

$\psi_R(P) = (i\epsilon)^2N + (i\epsilon)^4\frac{N^2}{2!}M + (i\epsilon)^6\frac{N^3}{3!}\frac{M^2}{2!} + \ldots$

$= -\epsilon\left(a - \frac{a^2}{2!}b + \frac{a^3}{3!}\frac{b^2}{2!}\ldots\right) = -\epsilon\frac{\partial}{\partial b}\left(ab - \frac{a^2b^2}{2!^2} + \frac{a^3b^3}{3!^2}\ldots\right)$

$= +\epsilon\frac{\partial}{\partial b}J_0(2\sqrt{ab})$ ← formally comes from derivative of J_0 at $b=0$ physically from no turn paths.

$= -\epsilon\sqrt{\frac{a}{b}}J_1(2\sqrt{ab}) + \epsilon\,\delta(b)$

To check: $\frac{\partial\psi_L}{\partial b} = i\psi_R$ is automatic. $\frac{\partial\psi_R}{\partial a} = -\epsilon\frac{\partial}{\partial a}\left[\sqrt{\frac{a}{b}}J_1(2\sqrt{ab})\right] = -\frac{\epsilon}{2b}\frac{\partial}{\partial a}\left[2\sqrt{ab}\,J_1(2\sqrt{ab})\right]$

$= -\frac{\epsilon}{2b}\frac{2}{2}\sqrt{\frac{b}{a}}\cdot 2\sqrt{ab}\,J_0(2\sqrt{ab}) = -\epsilon J_0(2\sqrt{ab}) = i\psi_L$ as required

The factor ϵ just becomes $\frac{dx}{2}$ on integrating green's function.

Hence if at $t = t_1$, $x = x_1$, ψ is $\psi_R(x_1,t_1)$ and ψ_L is zero, then at t_2, x_2

get ($\mu = 1$) $\psi_L(x_2,t_2) = i\int J_0\left(\sqrt{(t_2-t_1)^2 - (x_2-x_1)^2}\right)\psi_R(x_1,t_1)\frac{dx_1}{2}$ inside light cone

region of int.

$\psi_R(x_2,t_2) = -\int\frac{(t_2-t_1+x_2-x_1)}{\sqrt{}}J_1(\sqrt{})\psi_R(x_1,t_1)\frac{dx_1}{2} + \int\delta(t_2-t_1-x_2+x_1)\psi_R(x_1,t_1)\,dx_1$

Or in case both ψ_L and ψ_R are not zero initially obtain the complete greens formula: (OVER).

Fig. 7.1 Feynman's interpretation of the one-dimensional Dirac equation as a superposition of the zigzag paths of an electron (*Dirac Equation b*, folio 10, ca. 1946, same folio as in Fig. 4.2, reprinted with permission of Melanie Jackson Agency, LLC)

energy predicted by the Dirac equation.[2] Taking as his basis an idea that John A. Wheeler, his PhD supervisor, had communicated to him in the autumn of 1940, Feynman conceived of the positron as an electron moving backwards in time.[3] As a consequence, Feynman also had to consider paths that are directed backwards in time in his representation of the model, in particular paths that go twice through one point, which Feynman called "closed loop[s]" (see Fig. 7.2).

Feynman justified the Dirac equation by pointing out that it results from the superposition of the effects of the zigzag paths. Such a justification required that the number of possible paths could be determined and the contributions of all possible paths added to one another. The existence of closed loops made this impossible, at least at first sight: there were infinitely many paths and their contributions would not diminish in order to yield a finite value when added to one another. However, Feynman recognized that the existence of a path featuring a closed loop implies the existence of a path in which the closed loop is circumscribed in the opposite direction and that the contributions of the two paths would cancel each other out: "any completely closed loop cancels" (see Fig. 7.2, number 1). Therefore, it was not necessary to take into account the paths featuring closed loops and the procedure for solving the Dirac equation by counting paths was applicable, just as in the previously considered case of electrons and paths going only forwards in time.

Feynman proved the irrelevance of closed loops by diagrammatic induction (see Section 4.6.2 and Appendix A). Of course, he could not draw the abstract concept of what, in general, constitutes a closed loop; he could only draw instances of the "closed loop". However, Feynman's drawings show that every loop is always composed of the same elements.

As far as an individual element is concerned, Feynman knew that the sign of its contribution would change if the path is circumscribed in one or the other direction. Using some examples, Feynman tested his claim that the contributions of a closed loop that is circumscribed first in one direction and then in the other direction cancel each other out. Because in Feynman's drawings it is clear that the individual elements of a complex closed loop also yield contributions with the opposite sign (depending on the direction of the path), the claim proved to be valid for the general case, which is simply a composition of these individual elements (see Fig. 7.2 and 7.3, in particular number 2).

Feynman's next goals were to generalize his model of the quivering electron to more than one particle and also to describe their interactions. The Dirac equation takes interaction into account only indirectly: an electron generates an electromagnetic field, which acts on another electron. Again, Wheeler was influential in this respect: Feynman based his ideas on earlier work that he had done with Wheeler

---

[2] See, e. g., Dirac (1935, p. 271); cf. Dirac (1930).

[3] Schweber (1986a, p. 460); Schweber (1994, pp. 387/388); Feynman (1966, pp. 702, 705, 706); Weiner (1966, p. 32).

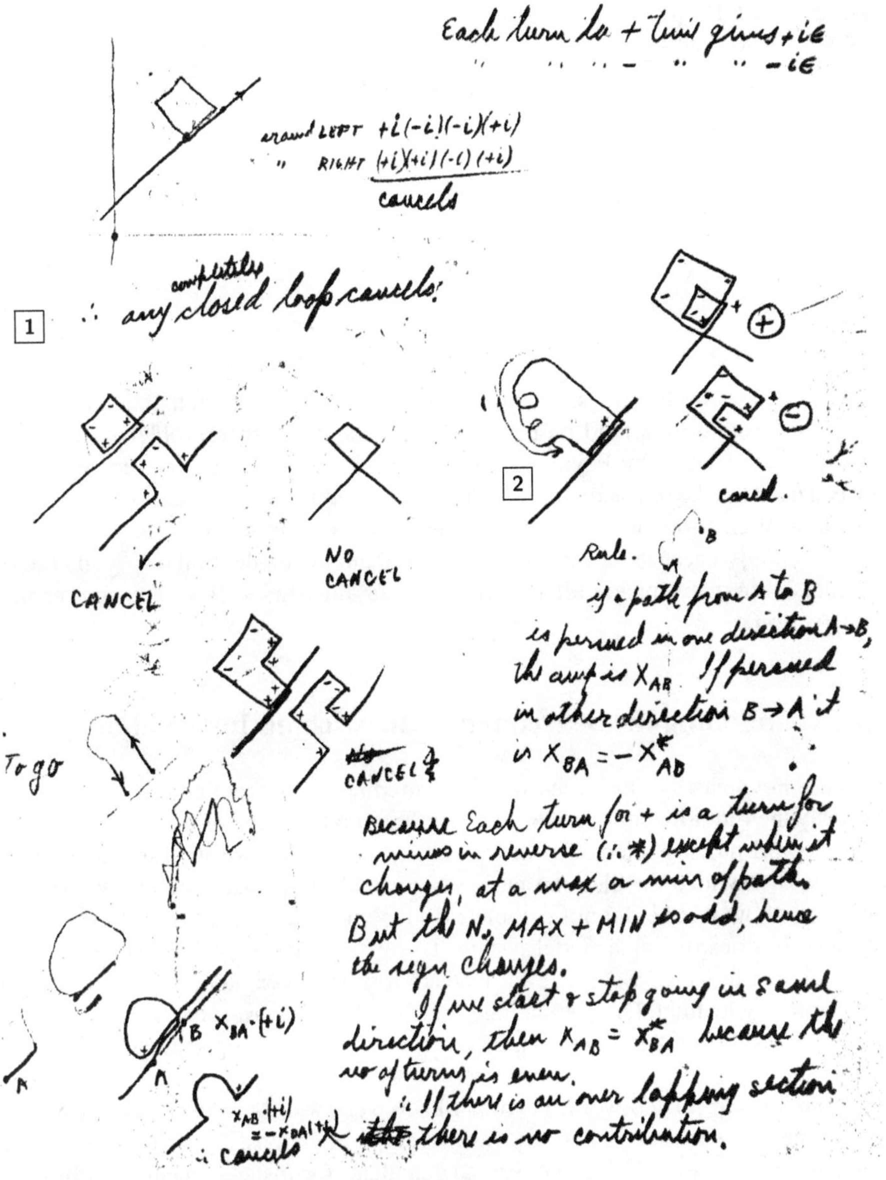

Fig. 7.2 Incorporation of positrons, which are represented as electrons going backwards in time, into the model of the quivering electron, as in the Dirac equation; and proof that closed loops do not have to be taken into account. (*Space-Time Approach to Quantum Electrodynamics*, folio 5. ca. 1947, same folio as in Fig. 4.19, reprinted with permission of Melanie Jackson Agency, LLC)

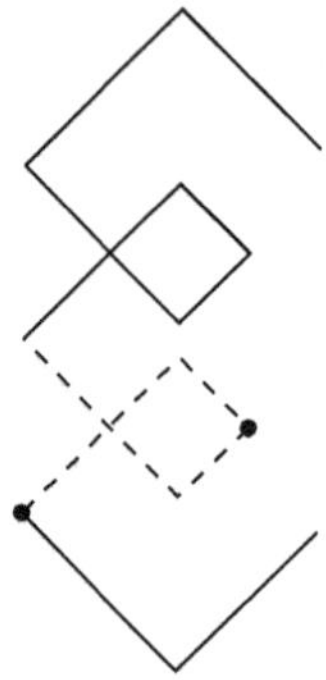

Fig. 7.3 A closed loop composed of other closed loops; identical to Fig. 4.24

on classical electrodynamics,[4] from which Feynman had learned that the mediation of the interaction by a field might account for the divergence problems. It is then easy to understand why Feynman gave up working with the Dirac equation and instead tried to take into account the interaction directly, just as he did when working on classical electrodynamics with Wheeler. However, he was dissatisfied with the results, mainly because he was only able to extend his model to the special case of an interaction that vanishes after a certain time; "this stinks" Feynman commented (see Section 4.7.2).

7.4 Abandoning the Search for a Microscopic Justification

Because Feynman was unsuccessful in his attempts to incorporate interaction satisfactorily, he did not continue working with the model of the quivering, backward- and forward-moving, electron, the movements of which are unobservable, since the area over which the electron moves is so small. Feynman's published articles, however, show that he did not completely give up his attempts to justify the Dirac equation by constructing a model system. The difference is only that he now pursued his goal on a less microscopic level. Instead of trying to explain the propagation of an electron, which the Dirac equation describes, by a superposition of microscopic zigzag paths, Feynman turned to a more abstract means of representation: Green's functions.

From his war-related work, Feynman knew very well how, using Green's functions, one can construct solutions to differential equations.[5] He now applied the methods that he used during the war, to calculate, for instance, neutron diffusion, to the Schrödinger and Dirac equations. In contrast to the treatment of the Dirac equation described above, Feynman no longer tried to give a more detailed account of the free propagation of an electron from one point to the other; this remained

[4] Cf. Wheeler and Feynman (1945, 1949).

[5] Cf. Galison (1998).

unanalyzed. Rather, by means of the Green's functions and the established method of perturbative expansion, he was able to represent the interaction of an electron in an electromagnetic field as a sequence of free propagations that are interrupted by infinitesimal amounts of disturbance by the external potential. In order to determine the total effect of the potential, Feynman integrated over all the space-time points where an infinitesimal interaction with the potential occurs (see Section 5.3).

Feynman's first published graphical representations were in the tradition of quantum theoretical wave mechanics: an incoming probability wave is scattered off a potential and produces a modified outgoing wave (see Fig. 7.4a). Even within this first publication (ThPos), the graphical representations become increasingly more abstract. In subsequent figures, the wave-mechanical elements are dropped, and simple lines represent the propagation of electrons and positrons. A circle or disc indicate the region in which the electromagnetic potential acts on the particles (see Fig. 7.4b). This figure also shows that Feynman was including the case of more than one particle in his considerations—though not yet with interactions. At this stage in the construction of his theory, Feynman assumed that free propagation is disturbed only by *external* potentials, not by potentials produced by other particles in the system under consideration.

It is only in his next publication (STQED) that Feynman succeeded in adequately (for him at any rate) describing the interaction between two electric particles. To obtain such a description he took as his basis the classical mathematical expression for the potential and reinterpreted it as a Green's function describing a quantum of the electromagnetic field: a photon. Feynman also adopted and interpreted innovatively not only the classical mathematical expression for the potential by which two electric particles interact but also its *graphical* representations. Wheeler and Feynman had used such representations in their alternative theory of electrodynamics (see Fig. 7.5a), and Feynman had also used such representations in his first attempts to incorporate the interaction between particles (see Fig. 7.6, number 1).

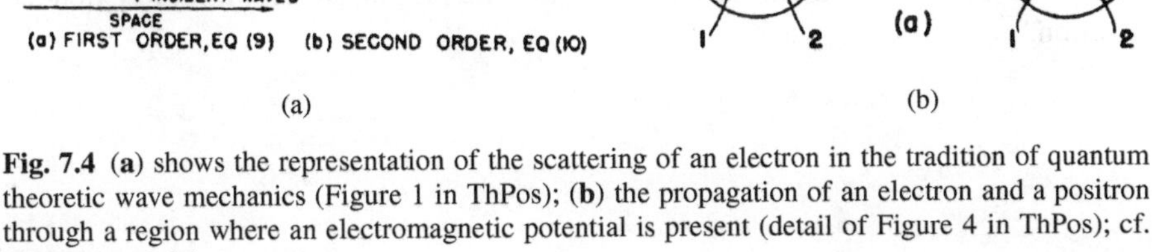

(a) (b)

Fig. 7.4 (**a**) shows the representation of the scattering of an electron in the tradition of quantum theoretic wave mechanics (Figure 1 in ThPos); (**b**) the propagation of an electron and a positron through a region where an electromagnetic potential is present (detail of Figure 4 in ThPos); cf. Figs. 5.1 and 5.4

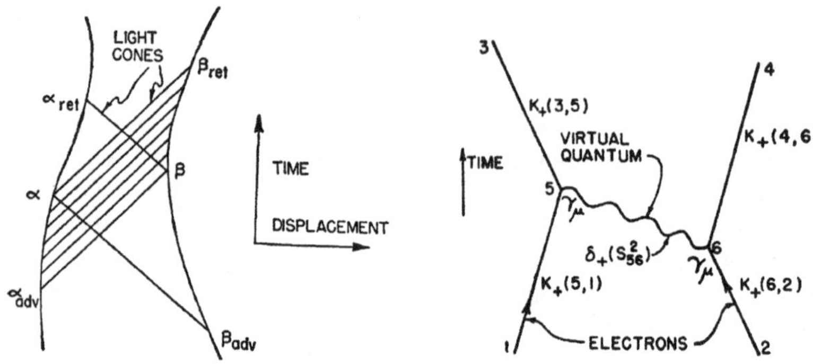

(a) (b)

Fig. 7.5 (**a**) shows the representation of the retarded and advanced potentials between two electrons according to the Wheeler–Feynman theory of electrodynamics (Figure 3 in Wheeler and Feynman 1949); (**b**) the fundamental interaction of QED (Figure 1 in STQED); see Figs. 4.30 and 5.8

7.5 The Successful Modification of the Theory

7.5.1 Feynman's Proposal

With the innovative interpretation of the classical interaction as a propagation of a light quantum, Feynman fulfilled his aspiration to find a "picture", the "slight modification" of which would remove the theory's problematic infinities. He had now reduced all QED processes to the propagation of electrons, positrons and photons. All QED processes are composed of the fundamental process shown in Fig. 7.5b. Feynman had thus found what he had been looking for ever since he began his "struggle" to understand fully the Dirac equation: a means of representation that makes explicit which features of the theory can be held responsible for its divergence problems. If the entire theory of QED is essentially made up of one fundamental process, an appropriate modification of this process should suffice to eliminate the problems. Again, Feynman's knowledge of similar classical cases was important: he knew from classical electrodynamics that the divergences could be avoided if the δ function, which represents the classical interaction potential and which is centred around one of its arguments, was replaced with a more extended function.[6]

[6] Bopp 1940, 1942, McManus 1948, cf. see also footnote 4 in CutOffCl.

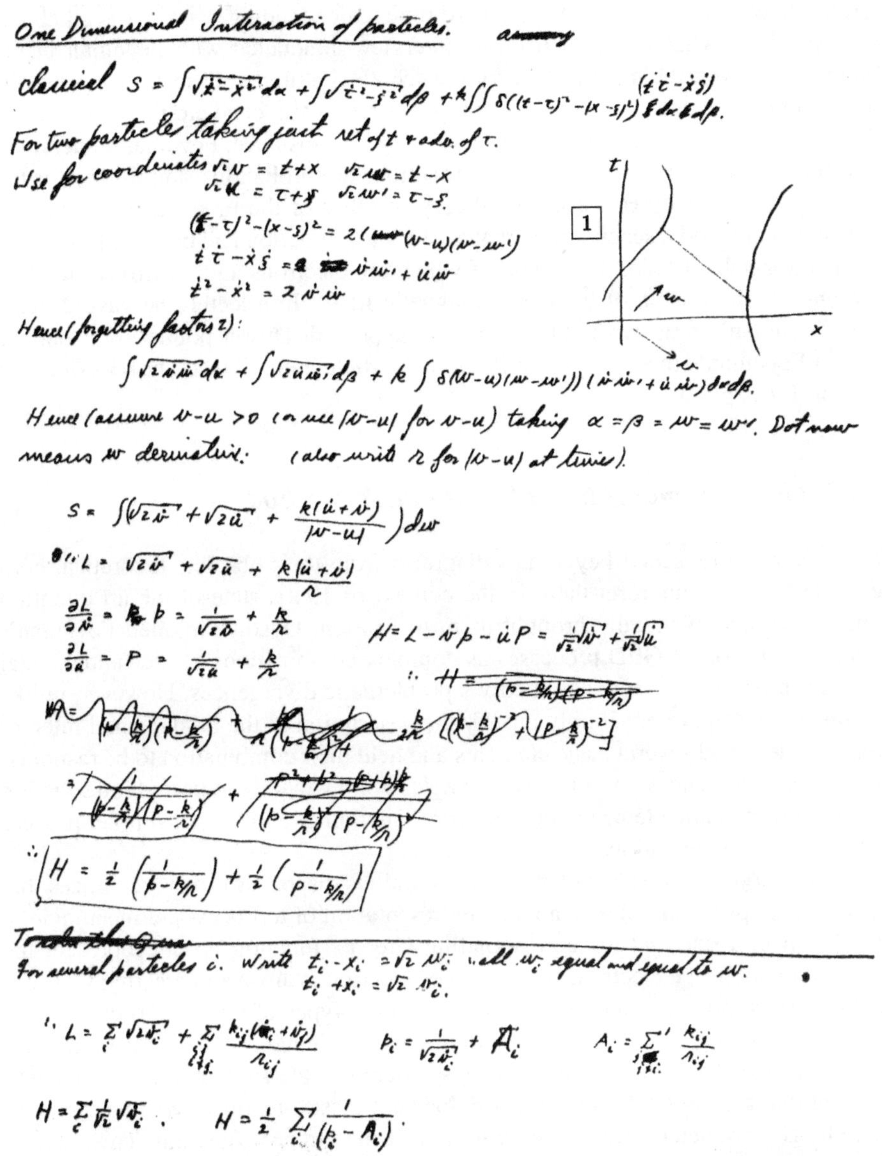

Fig. 7.6 The classical interaction potential between two particles in one spatial dimension (*Dirac Equation h*, folio 5, ca. 1947, same folio as in Fig. 4.29, reprinted with permission of Melanie Jackson Agency, LLC)

7.5.2 Critique

The community of physicists working at the time received Feynman's modified theory of QED, with which Feynman intended to solve the problems of the then current theory, sceptically and had a hard time understanding what it was all about; at least this is what we learn from an interview conducted with Feynman in the 1960s.[7] However, the main reason for the sceptical attitude of most physicists was not Feynman's extensive use of diagrams but rather the obsolete theoretical principles on which Feynman had based his theory. For instance, Feynman understood that, like in quantum mechanics, wave functions are probability amplitudes for the position of a particle. However, the theory of QED of the time was a theory that had been quantized a second time, that is, the wave function had been replaced with an operator-valued field, the quanta of which are electrons and positrons (just like photons are the quanta of the electromagnetic field). In a lecture he gave 2 years after Feynman's first two publications had appeared, Dyson pointed out that the flaw in Feynman's theory was that it was a "particle theory" (see Dyson's quote in Section 1.4.1, p. 10).

7.5.3 The Systematization of Feynman's Method

It is Dyson who rescued Feynman's diagrams from their obsolete theoretical context. He began to interpret them in the context of 1940s state-of-the art quantum field theory and eventually brought them to fruition. Dyson exploited Feynman's representation of all QED processes as complex combinations of the fundamental element in order to remove the theory's problematic divergences. However, unlike Feynman, Dyson did not simply modify the description of the fundamental interaction. He identified several basic elements and held their combination to be responsible for the divergences. Dyson was then able to show that the repeated introduction of these problematic elements into the representation of a given but simpler process led to the infinite quantities.

Furthermore, Dyson showed that these infinities disappear if one recognizes that the elements of the simplest graphical representation of a process are meaningless. For instance, a line and the corresponding Green's function do not represent the propagation of a free electron, since a "free" electron is always interacting with virtual electrons, positrons and photons or even other types of virtual particles. These virtual processes are represented by certain types of elements that are inserted into a basic Feynman diagram (see Fig. 7.7). Dyson demonstrated that, in order to take into account that a completely free electron does not exist, one must redefine the mass and charge parameter used in the definition of the Green's function. This method became known as *renormalization*.

Through mass and charge renormalization, the inert, that is force-free, state of a particle is redefined. For instance, one defines the energy of a free electron as the observed (finite) energy of an electron, which necessarily includes the virtual processes, such that the electron is not free in the strict sense of the word. Through the renormalization of the mass and charge, the uninterpretable infinities are absorbed

[7] Weiner (1966); see also Schweber 1994.

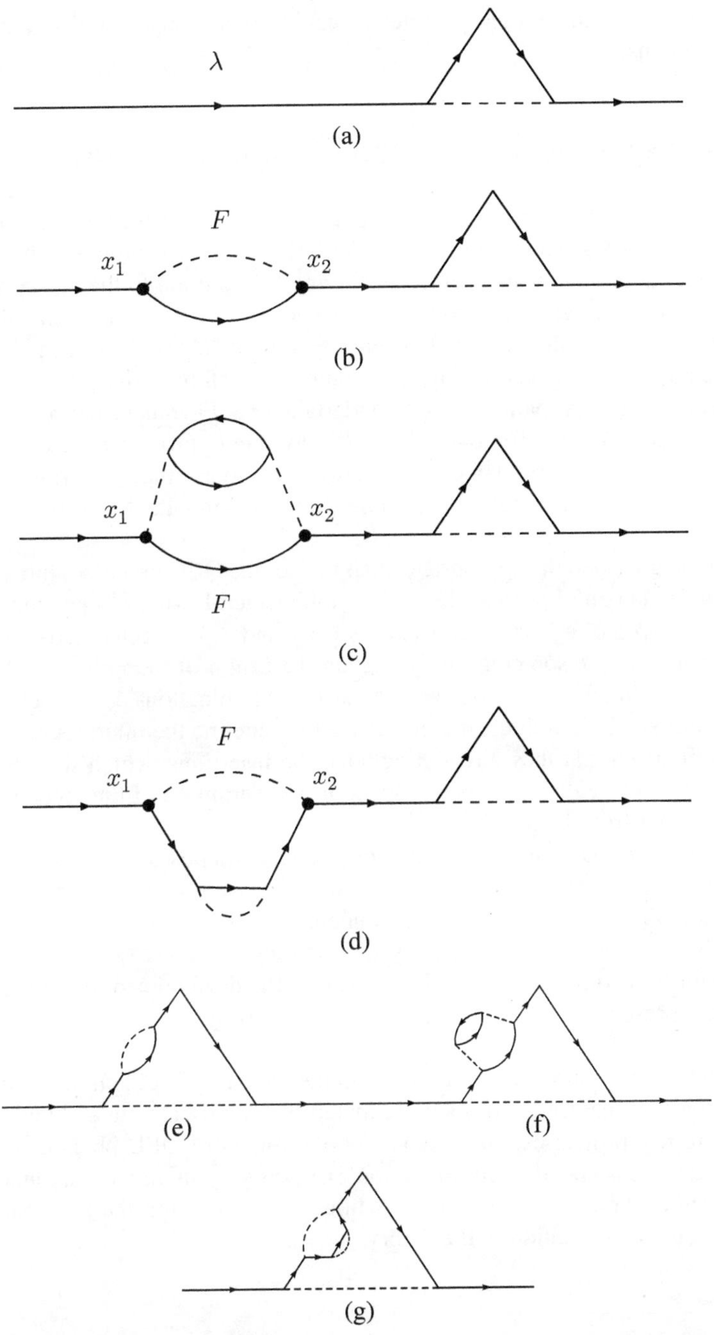

Fig. 7.7 Insertion into the basic diagram (**a**) of elements that are responsible for the theory's divergences; identical to Fig. 6.3.

into unobservable theoretical parameters, and thus no longer pose any interpretational problems.

7.6 The Explanatory Power of the New Means of Representation

Both Feynman and Dyson maintained that a large part of their work was a mere reformulation of known theoretical content.[8] However, reformulating the problematic theoretical content was essential to finding the solution to the theory's theoretical problems. The representation of QED phenomena that was eventually established by Dyson revealed that the concept of a completely free electron is meaningless. Once this was taken seriously, the problematic infinities disappeared.

Drawing on the preparatory work undertaken by Feynman, but also by Julian Schwinger and Sin-Itiro Tomonaga, Dyson was able to provide an adequate model of QED's well-known fundamental equations, a model, that is, "from which the well-known divergences seem to have conspired to eliminate themselves" (see Section 6.6.2).

Feynman provided the appropriate "parsing" of the QED processes into combinations of a fundamental process. Dyson complemented Feynman's parsing by introducing new types of elements: self-energy parts and disconnected parts. With these elements in hand, Dyson could then explain the origin of the infinities: they arose from the fact that QED processes are fractal-like combinations of these elements. In order to remove the infinities, it then sufficed to redefine the interpretation of these basic elements so that they already included the insertions, which were repeatable *ad infinitum*. Physically, this corresponded to a redefinition of the theoretical mass and charge of a free electron.

The whole of Dyson's analysis, however, presupposes a specific and non-traditional representation of QED processes. When one is using Feynman diagrams, one implicitly claims that the QED phenomena have a certain structure. Only on the basis of this specific structure could Dyson give reasons for the existence of infinities and explain how to remove them. In that sense, the development of an appropriate means of representation was part and parcel of the solution to the divergence problems.

Dyson did not subtract the infinities from the theoretical predictions with the sole justification that the predictions should match the empirical data and be consistent with theoretical principles. To some extent, the model of QED phenomena that he provided using his specific representation *explains* why there were uninterpretable divergences in the unmodified theory, while also explicating the physical content inherent in the modification of the theory.

---

[8] For example, "simply rewriting expressions in a simpler form" (STQED, p. 776); "using no new ideas" (SM, p. 1754); "[Tomonaga, Schwinger, and Feynman] kept the physical basis of the theory precisely as it had been laid down by Dirac, and only changed the mathematical superstructure" (Dyson 1965, p. 589).

It remains to be investigated in more detail to what extent Feynman diagrams are an instance of a pattern that can be generalized to other cases in the history of science in which problems are not solved in the usual sense of the word but are rather made to disappear by using a symbol system that appropriately represents an adequate model. I surmise that it should be possible to show, more explicitly than I have done in this dissertation, that, using the symbol system of old QED, one can express what could be seen as the force-free motion of a completely isolated electron. As no physical system corresponds to such an expression, one could then conclude that the symbol system of the new theory of QED is superior to the old symbol system because it does not allow for such an expression and thus clearly reveals that the concept of a completely isolated electron is meaningless. And because the meaningless concept was eliminated through the use of the new means of representation, the disappearance of the divergence problems, to which it gave rise, was, as Dyson wrote, "automatic".

Appendix A
Diagrammatic Induction

It might seem as though one of the limitations of diagrammatic proofs is that they can only express specific statements, not general ones. If one wants to prove that the sum of all the angles in a triangle is 180°, it does not suffice to draw one triangle and to confirm (approximately) the statement. So how can one prove or even express a general statement using a diagram?

One possibility is by *inductive* proof, much in the same way that inductive proofs are carried out using mathematical formulae. An inductive proof does not need the special features of mathematical or logical formulae and can therefore be performed using other types of symbols as well. In fact, it is even not obvious how one should distinguish mathematical or logical formulae from other types of symbols such as diagrams (see, e. g., Goodman 1968, chapter IV, section 10).

In logic and mathematics one can, for example, prove by induction that a statement is generally true when it has been explicitly verified only for a finite number of special cases. This is possible by establishing a prescription of how all the special cases, one after the other, can be constructed, such that at each stage of the construction the truth of the statement is preserved. If such a prescription is available, the same inductive proof procedure should, therefore, be possible for any kind of symbol, including symbols that one calls diagrams, graphs or drawings. For example, the statement that the sum of the odd numbers up to $2n - 1$ equals n^2 can be proved by induction using either mathematical formulae *or* diagrams. Brown (1996, p. 253) also discusses this example; note, however, that he does *not* consider that the variant of the proof using diagrams is a proof by induction.

The standard proof using mathematical formulae is first to verify the statement for the first $n = 1$:

$$\sum_{j}^{1} 2j - 1 = 1 = 1^2. \tag{A.1}$$

A. Wüthrich, *The Genesis of Feynman Diagrams*, Archimedes 26,
DOI 10.1007/978-90-481-9228-1, © Springer Science+Business Media B.V. 2010

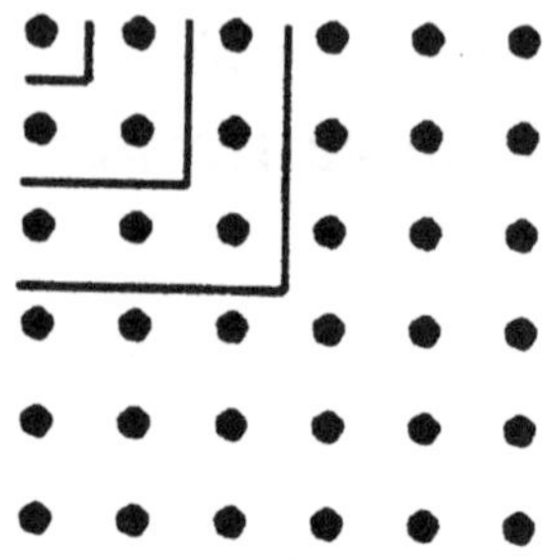

Fig. A.1 Diagram used to prove that the sum of odd numbers up to $2n - 1$ equals n^2 (Brown 1996, p. 251)

Next, one expresses the general sum up to $2n - 1$ in terms of the sum with one term less, that is up to $2(n - 1) - 1$:

$$\sum_{j}^{n} 2j - 1 = \sum_{j}^{n-1} (2j - 1) + 2n - 1. \tag{A.2}$$

Assuming that the statement holds for the sum with one term less, one obtains

$$(n - 1)^2 + 2n - 1 = n^2 - 2n + 1 + 2n - 1 = n^2. \tag{A.3}$$

Thus one has proved that if the statement holds for a sum up to a certain n, it also holds for the sum with one more term. Since the statement was verified for the lowest possible n, it holds for any n.

Therefore, proof by induction essentially hinges on one being able to construct one member of the class of objects about which one wants to derive a statement out of the previous member, relative to a certain ordering of the members in the class. Yet this can also be done using a *graphical* representation of the members of the class, in our case the odd numbers up to $2n - 1$. One can represent the sequence of odd numbers as sectors in a quadratic array of dots containing n^2 dots (see Fig. A.1). Then one can verify the statement $\sum_{j}^{n} 2j - 1 = n^2$ for the special case $n = 1$: one dot is the same as a quadratic array made up of one dot. The next member of the class is obtained by adding an L-shaped segment to the previous member in the lower right-hand side of the diagram, since the L-shaped segment contains an odd number of dots. Adding an L-shaped segment to a square results again in a square, the side of which contains one more dot than the original square. Therefore, if the statement is true for the previous member, it is true for the next member. Since any member of the class can be constructed by adding L-shaped segments, the statement is proved in its full generality.

Appendix B
Synopsis of Manuscripts and Principal Publications

On the following page, I give an overview of the notes and the publications by Feynman as well as the publications by Freeman J. Dyson, Hans Euler and the Japanese physicists working around Sin-Itiro Tomonaga that I discuss in my dissertation. I also include some of the publications by Julian Schwinger and Ernst C. G. Stueckelberg that might be of interest as regards the chronology of the development of Feynman diagrams. For the published work, I give the date of publication; for Feynman's notes, I give the approximate dates that seem most plausible to me on the basis of my reconstruction of the developments.

Manuscripts by Feynman		Publications by Feynman		Publications by others	
				1934	Stueckelberg
				1936	Euler
				1938	Stueckelberg
		1941	Wheeler and Feynman (conference presentation, see 'Minutes of the . . .' 1941)	1941	Stueckelberg
		1942	PhD thesis (Feynman 2005)		
		1945	Wheeler and Feynman		
ca. 1946	*Dirac Equation b*, *Harmonic Oscillators b*			1946	Tomonaga
1946/47	Exchange of letters between Feynman and Theodore Welton (*Dirac Equation a*)				
ca. 1947	*Space-time approach to quantum electrodynamics*, *Dirac Equation h*				
ca. 1948	*Advanced Quantum Mechanics (at Cornell)*, *Theory of Positrons . . .*	1948	RMP48 (April), CutOffCl (Oct.), CutOffQ (Nov.)	1948	Schwinger 1948a (Feb.), Schwinger 1948b (Nov.), RadReacI, RadReacII
		1949	Wheeler and Feynman 1949 (July), ThPos (Sept.), STQED (Sept.)	1949	RadReacIIIa, RadReacIIIb, RadTh (Feb.), SM (June)
early 1950s	*Helium*	1951	Feynman		

References

Abbreviations

RMP48	R. P. Feynman (1948). 'Space-Time Approach to Non-Relativistic Quantum Mechanics'. In: *Reviews of Modern Physics* 20.2 (Apr. 1948), pp. 367–387.
ThPos	R. P. Feynman (1949). 'The Theory of Positrons'. In: *Physical Review* 76.6 (Sept. 1949), pp. 749–759.
STQED	R. P. Feynman (1949). 'Space-Time Approach to Quantum Electrodynamics'. In: *Physical Review* 76.6 (Sept. 1949), pp. 769–789.
RadTh	F. J. Dyson (1949). 'The Radiation Theories of Tomonaga, Schwinger, and Feynman'. In: *Physical Review* 75.3 (Feb. 1949), pp. 486–502.
CutOffCl	R. P. Feynman (1948). 'A Relativistic Cut-Off for Classical Electrodynamics'. In: *Physical Review* 74.8 (Oct. 1948), pp. 939–946.
CutOffQ	R. P. Feynman (1948). 'Relativistic Cut-Off for Quantum Electrodynamics'. In: *Physical Review* 74.10 (Nov. 1948), pp. 1430–1438.
SM	F. J. Dyson (1949). 'The S Matrix in Quantum Electrodynamics'. In: *Physical Review* 75.11 (June 1949), pp. 1736–1755.
RadReacI	Z. Koba and S.-I. Tomonaga (1948). 'On Radiation Reactions in Collision Processes. I.' In: *Progress of Theoretical Physics* 3.3, pp. 290–303.
RadReacII	Z. Koba and G. Takeda (1948). 'Radiation Reaction in Collision Process, II'. In: *Progress of Theoretical Physics* 3.4, pp. 407–421.
RadReacIIIa	Z. Koba and G. Takeda (1949). 'Radiation Reaction in Collision Process III1. First Radiative Correction for an Arbitrary Process Including Electrons, Positrons, and Light Quanta'. In: *Progress of Theoretical Physics* 4.1, pp. 60–70.
RadReacIIIb	Z. Koba and G. Takeda (1949). 'Radiation Reaction in Collision Process III2. First Radiative Correction for an Arbitrary Process Including Electrons, Positrons, and Light Quanta'. In: *Progress of Theoretical Physics* 4.2, pp. 130–141.

Cross-References

The bibliography partially functions as an index: At the end of each entry the pages on which the item is cited are indicated by the phrase "See p./pp. ... ".

Classification of the Material from the Feynman Collection Held by the Caltech Archives

The Papers of Richard Phillips Feynman held by the Caltech (California Institute of Technology) Archives have been filed into labelled folders and boxes (91 in total). A finding aid is available at the Archives and online.[1] The Feynman Collection was processed by Charlotte E. Erwin, Carol Finerman and David A. Valone and completed on 1 July 1993. The finding aid was updated in September 2002. The Archives kindly permitted me to take digital photographs of parts of those folders which contained possibly relevant material for my area of investigation. Thus, from some of the folders I selected several series of folios, which I labelled, for the purposes of my research, *a*, *b*, *c*, etc., in the order that they appear in the folder. The particular folios in a series are referred to as the first, second, third, etc., folio in the series in the order that they appear in it, irrespective of any page-numbering system that Feynman may have applied to his notes. Note that under normal circumstances the folios made available to researchers are photocopies of Feynman's original papers. Therefore, Feynman's notes on the verso of a piece of paper appear as separate folios in the Archives. I have used the title given by the archivists to refer to a particular folder in a box, the numbers of which are given in the references below. For example, "*Dirac Equation a*, folio 11" refers to the 11th folio in series *a* from Folder 2 in Box 11, which was given the title "Dirac Equation" by the archivists. If no series label is given, then I have selected only one series of folios from the relevant folder (but the whole folder is not necessarily made up of only this one series).

I was able to date only one of the folios precisely (*Dirac Equation a*, folio 1, see page 80). Nevertheless, I indicate for each series of folios the approximate dates that seem most plausible to me on the basis of my reconstruction of the developments.

For internal use only, I have stored my selection of 588 folios in `jpeg` format in the digital library of the Institute of Philosophy of the University of Bern. On request, and with the agreement of the Caltech Archives, the Institute of Philosophy will grant interested researchers access to this material.

[1] The online finding aid is provided by the Online Archive of California, and is available at `http://www.oac.cdlib.org/findaid/ark:/13030/kt5n39p6k0`, last visited 12 July 2010.

References to the Material from the Feynman Collection

Advanced Quantum Mechanics (at Cornell). (ca. 1948). Series of selected folios from Box 7, Folder 4.

Dirac Equation a. (1946/47). First series of selected folios from Box 11, Folder 2.

Dirac Equation b. (Ca. 1946). Second series of selected folios from Box 11, Folder 2.

Dirac Equation h. (Ca. 1947). Eighth series of selected folios from Box 11, Folder 2.

Theory of Positrons; Self-Energy in an Atom; Radiation Scattering; Spreading Dirac Packet. (Ca. 1948). Series of selected folios from Box 13, Folder 1.

Harmonic Oscillators b. (Ca. 1946). Second series of selected folios from Box 11, Folder 5.

Helium. (Early 1950s). Series of selected folios from Box 11, Folder 6.

Space-Time Approach to Quantum Electrodynamics. (Ca. 1947). Series of selected folios from Box 12, Folder 9.

Other References

Baeyer, H. C. von (1999). 'Nota Bene'. In: *The Sciences* (January/February 1999). New York, NY: The New York Academy of Sciences, pp. 12–15.

Bethe, H. A. (1947). 'The Electromagnetic Shift of Energy Levels'. In: *Physical Review* 72.4 (Aug. 1947), pp. 339–341.

Bethe, H. A. and S. S. Schweber (1955). *Mesons and Fields*. New York, NY: Row, Peterson.

Bopp, F. (1940). 'Eine lineare Theorie des Elektrons'. In: *Annalen der Physik* 38, pp. 345–384.

Bopp, F. (1942). 'Lineare Theorie des Elektrons. II'. In: *Annalen der Physik* 42, pp. 573–608.

Breit, G. (1928). 'An Interpretation of Dirac's Theory of the Electron'. In: *Proceedings of the National Academy of Sciences of the United States of America* 14.7, pp. 553–559.

Breit, G. and J. A. Wheeler (1934). 'Collision of Two Light Quanta'. In: *Physical Review* 46.12 (Dec. 1934), pp. 1087–1091.

Brown, H. R. and R. Harrè, eds. (1988). *Philosophical Foundations of Quantum Field Theory*. Oxford: Clarendon Press.

Brown, J. R. (1996). 'Illustration and Inference'. In: *Picturing Knowledge: Historical and Philosophical Problems Concerning the Use of Art in Science*. Ed. by B. Baigrie. Toronto, ON: University of Toronto Press. Chap. 8, pp. 250–268.

Brown, L. M., ed. (1993). *Renormalization. From Lorentz to Landau (and Beyond)*. New York, NY: Springer.

Carson, C. (1996a). 'The Peculiar Notion of Exchange Forces–I: Origins in Quantum Mechanics, 1926–1928'. In: *Studies In History and Philosophy of Science Part B: Studies In History and Philosophy of Modern Physics* 27.1 (Mar. 1996), pp. 23–45.

Carson, C. (1996b). 'The Peculiar Notion of Exchange Forces–II: From Nuclear Forces to QED, 1929–1950'. In: *Studies In History and Philosophy of Science Part B: Studies In History and Philosophy of Modern Physics* 27.2 (June 1996), pp. 99–131.

Cartwright, N. (1983). *How the Laws of Physics Lie*. Oxford: Clarendon Press.

Compton, A. H. (1923). 'A Quantum Theory of the Scattering of X-rays by Light Elements'. In: *Physical Review* 21.5 (May 1923), pp. 483–502.

Dancoff, S. M. (1939). 'On Radiative Corrections for Electron Scattering'. In: *Physical Review* 55.10 (May 1939), pp. 959–963.

Devoto, A. and D. Duke (1984). 'Table of integrals and formulae for Feynman diagram calculations'. In: *La Rivista del Nuovo Cimento (1978–1999)* 7.6 (June 1984), pp. 1–39.

Dirac, P. A. M. (1930). 'A Theory of Electrons and Protons'. In: *Proceedings of the Royal Society of London. Series A, Containing Papers of a Mathematical and Physical Character* 126.801, pp. 360–365.

Dirac, P. A. M. (1933). 'The Lagrangian in Quantum Mechanics'. In: *Physikalische Zeitschrift der Sowjetunion* 3.1, pp. 64–72.

Dirac, P. A. M. (1935). *The Principles of Quantum Mechanics by P. A. M. Dirac*. 2nd ed. Oxford: Clarendon Press.

Dirac, P. A. M. (1938). 'Classical Theory of Radiating Electrons'. In: *Proceedings of the Royal Society of London. Series A, Mathematical and Physical Sciences* 167.929 (Aug. 1938), pp. 148–169.

Dirac, P. A. M. (1946). 'Elementary Particles and their Interactions'. In: *The Collected Works of P. A. M. Dirac, 1924–1948*. Ed. by R. H. Dalitz. Paper given at Princeton University Bicentennial. Cambridge: Cambridge University Press.

Dresden, M. (1993). 'Renormalization in Historical Perspective—The First Stage'. In: *Renormalization. From Lorentz to Landau (and Beyond)*. Ed. by L. M. Brown. New York, NY: Springer.

Dyson, F. J. (1949). 'The Radiation Theories of Tomonaga, Schwinger, and Feynman'. In: *Physical Review* 75.3 (Feb. 1949), pp. 486–502.

Dyson, F. J. (1949). 'The S Matrix in Quantum Electrodynamics'. In: *Physical Review* 75.11 (June 1949), pp. 1736–1755.

Dyson, F. J. (1965). 'Tomonaga, Schwinger, and Feynman Awarded Nobel Prize for Physics'. In: *Science*. 3rd ser. 150.3696 (Oct. 1965), pp. 588–589.

Dyson, F. J. (2006). *Advanced Quantum Mechanics: Second Edition*. URL: http://arxiv.org/abs/quant-ph/0608140v1.

Euler, H. (1936). 'Über die Streuung von Licht an Licht nach der Diracschen Theorie'. In: *Annalen der Physik* 418.5, pp. 398–448.

Euler, H. and B. Kockel (1935). 'Über die Streuung von Licht an Licht nach der Diracschen Theorie'. In: *Naturwissenschaften* 23.15 (Apr. 1935), pp. 246–247.

Fermi, E. (1932). 'Quantum Theory of Radiation'. In: *Reviews of Modern Physics* 4.1 (Jan. 1932), pp. 87–132.

Feynman, R. P. (1948). 'A Relativistic Cut-Off for Classical Electrodynamics'. In: *Physical Review* 74.8 (Oct. 1948), pp. 939–946.

Feynman, R. P. (1948). 'Relativistic Cut-Off for Quantum Electrodynamics'. In: Physical Review 74.10 (Nov. 1948), pp. 1430–1438.

Feynman, R. P. (1948). 'Space-Time Approach to Non-Relativistic Quantum Mechanics'. In: *Reviews of Modern Physics* 20.2 (Apr. 1948), pp. 367–387.

Feynman, R. P. (1949). 'Space-Time Approach to Quantum Electrodynamics'. In: *Physical Review* 76.6 (Sept. 1949), pp. 769–789.

Feynman, R. P. (1949). 'The Theory of Positrons'. In: *Physical Review* 76.6 (Sept. 1949), pp. 749–759.

Feynman, R. P. (1951). 'An Operator Calculus Having Applications in Quantum Electrodynamics'. In: *Physical Review* 84.1 (Oct. 1951), pp. 108–128.

Feynman, R. P. (1966). 'The Development of the Space-Time View of Quantum Electrodynamics'. In: *Science* 153.3737. Nobel lecture, pp. 699–708.

Feynman, R. P. (2005). 'The Principle of Least Action in Quantum Mechanics'. In: *Feynman's Thesis: A New Approach to Quantum Theory*. Ed. by L. M. Brown. Singapore: World Scientific Publishing, pp. 1–69.

Feynman, R. P. and A. R. Hibbs (1965). *Quantum Mechanics and Path Integrals*. New York, NY: McGraw-Hill.

Fokker, A. D. (1929). 'Ein invarianter Variationssatz für die Bewegung mehrerer elektrischer Massenteilchen'. In: *Zeitschrift für Physik* 58, pp. 386–393.

Frenkel, J. (1925). 'Zur Elektrodynamik punktförmiger Elektronen'. In: *Zeitschrift für Physik A Hadrons and Nuclei* 32.1 (Dec. 1925), pp. 518–534.

Frisch, M. (2005). *Inconsistency, Asymmetry, and Non-Locality: A Philosophical Investigation of Classical Electrodynamics*. Oxford: Oxford University Press.

Furry, W. H. (1937). 'A Symmetry Theorem in the Positron Theory'. In: *Physical Review* 51.2 (Jan. 1937), pp. 125–129.

Galison, P. (1983). 'The Discovery of The Muon and the Failed Revolution Against Quantum Electrodynamics'. In: *Centaurus* 26, pp. 262–316.

Galison, P. (1998). 'Feynman's War: Modelling Weapons, Modelling Nature'. In: *Studies In History and Philosophy of Science Part B: Studies In History and Philosophy of Modern Physics* 29.3 (Sept. 1998), pp. 391–434.

Giere, R. N. (1996). 'Visual Models and Scientific Judgement'. In: *Picturing Knowledge: Historical and Philosophical Problems Concerning the Use of Art in Science*. Ed. by B. Baigrie. Toronto, ON University of Toronto Press, pp. 267–302.

Gleick, J. (1992). *Genius: The Life and Science of Richard Feynman*. Originally published: New York: Pantheon Books/Vintage Books.

Goodman, N. (1968). *Languages of Art: An Approach to a Theory of Symbols*. Indianapolis, IN: Hackett Publishing.

Graβhoff, G., H.-C. Liess and K. Nickelsen (2002). 'Visualisiertes Wissen: Eine Einführung in die Analyse wissenschaftlicher Abbildungen'. Unpublished manuscript. History and Philosophy of Science. University of Bern.

Griffiths, D. (1987). *Introduction to Elementary Particles*. Hoboken, NJ: Wiley.

Halpern, O. (1933). 'Scattering Processes Produced by Electrons in Negative Energy States'. In: *Physical Review* 44.10 (Nov. 1933), pp. 855–856.

Harré, R. (1988). 'Parsing the Amplitudes'. In: *Philosophical Foundations of Quantum Field Theory*. Ed. by H. R. Brown and R. Harré. Oxford: Clarendon Press, pp. 59–71.

Heisenberg, W. (1934). 'Bemerkungen zur Diracschen Theorie des Positrons'. In: *Zeitschrift für Physik A Hadrons and Nuclei* 90.3 (Mar. 1934), pp. 209–231.

Heisenberg, W. (1943). 'Die "beobachtbaren Gröβen" in der Theorie der Elementarteilchen'. In: *Zeitschrift für Physik A Hadrons and Nuclei* 120.7 (July 1943), pp. 513–538.

Heitler, W. (1944). *Quantum Theory of Radiation*. 2nd ed. London: Oxford University Press.

Huggett, N. (2000). 'Philosophical Foundations of Quantum Field Theory'. In: *The British Journal for the Philosophy of Science* 51.4, pp. 617–637.

Jacobson, T. and L. S. Schulman (1984). 'Quantum Stochastics: The Passage from a Relativistic to a Non-relativistic Path Integral'. In: *Journal of Physics A: Mathematical and General* 17.2, pp. 375–383.

Kaempffer, F. A. (1965). *Concepts in Quantum Mechanics*. New York, NY: Academic.

Kaiser, D. (2005). *Drawing Theories Apart: The Dispersion of Feynman Diagrams in Postwar Physics*. Chicago: The University of Chicago Press.

Karplus, R. and M. Neuman (1950). 'Non-Linear Interactions Between Electromagnetic Fields'. In: *Physical Review* 80.3 (Nov. 1950), pp. 380–385.

Kauffman, L. H. and H. P. Noyes (1996). 'Discrete Physics and the Dirac Equation'. In: *Physics Letters A* 218.3–6 (Aug. 1996), pp. 139–146.

Koba, Z. and G. Takeda (1948). 'Radiation Reaction in Collision Process, II'. In: *Progress of Theoretical Physics* 3.4, pp. 407–421.

Koba, Z. and G. Takeda (1949a). 'Radiation Reaction in Collision Process III1. First Radiative Correction for an Arbitrary Process Including Electrons, Positrons, and Light Quanta'. In: *Progress of Theoretical Physics* 4.1, pp. 60–70.

Koba, Z. and G. Takeda (1949b). 'Radiation Reaction in Collision Process III2. First Radiative Correction for an Arbitrary Process Including Electrons, Positrons, and Light Quanta'. In: *Progress of Theoretical Physics* 4.2, pp. 130–141.

Koba, Z. and S.-I. Tomonaga (1948). 'On Radiation Reactions in Collision Processes. I.' In: *Progress of Theoretical Physics* 3.3, pp. 290–303.

Kuhn, T. S. (1962). *The Structure of Scientific Revolutions*. 1st ed. Vol. 2. International Encyclopedia of Unified Science 2. (2nd ed. 1970). Chicago: University of Chicago Press.

Kusch, P. and H. M. Foley (1948). 'The Magnetic Moment of the Electron'. In: *Physical Review* 74.3 (Aug. 1948), pp. 250–263.

Lacki, J., H. Ruegg and V. Telegdi (1999). *The Road to Stueckelberg's Covariant Perturbation Theory as Illustrated by Successive Treatments of Compton Scattering*. URL: http://arxiv.org/abs/physics/9903023.

Lamb, W. E. and R. C. Retherford (1947). 'Fine Structure of the Hydrogen Atom by a Microwave Method'. In: *Physical Review* 72.3 (Aug. 1947), pp. 241–243.

Leutwyler, H. (1986). 'Elektrodynamik und Optik'. Lecture notes, Institute for Theoretical Physics, University of Bern.

'Minutes of the Cambridge, Massachusetts, Meeting, February 21 and 22, 1941' (1941). In: Physical Review 59.8 (Apr. 1941), pp. 682–691.

McManus, H. (1948). 'Classical Electrodynamics without Singularities'. In: *Proceedings of the Royal Society of London. Series A, Mathematical and Physical Sciences* 195.1042, pp. 323–336.

Mehra, J. (1994). *The Beat of a Different Drum: The Life and Science of Richard Feynman*. New York, NY: Oxford University Press.

Meynell, L. (2008). 'Why Feynman Diagrams Represent'. In: *International Studies in the Philosophy of Science* 22.1, pp. 39–59.

Miller, A. I. (1984). *Imagery in Scientific Thought: Creating 20th-Century Physics*. Boston, MA: Birkhäuser.

Miller, A. I. (1995). *Early Quantum Electrodynamics: A Sourcebook*. Cambridge: Cambridge University Press.

Mills, R. (1993). 'Tutorial on Infinities in QED'. In: *Renormalization. From Lorentz to Landau (and Beyond)*. Ed. by L. M. Brown. New York: Springer.

Møller, C. (1931). 'Über den Stoβ zweier Teilchen unter Berücksichtigung der Retardation der Kräfte'. In: *Zeitschrift für Physik A Hadrons and Nuclei* 70.11 (Nov. 1931), pp. 786–795.

Møller, C. (1932). 'Zur Theorie des Durchgangs schneller Elektronen durch Materie'. In: *Annalen der Physik* 406.5, pp. 531–585.

Nickelsen, K. (2002). 'Entstehung, Inhalt und Funktion wissenschaftlicher Pflanzenbilder des 18. Jahrhunderts'. PhD thesis. University of Bern.

Oppenheimer, J. R. (1950). 'Electron Theory'. In: *Rapports du 8e Conseil de Physique, Solvay*. Reprinted in Schwinger 1958, pp. 145–155, 269–281.

Pais, A. (1986). *Inward Bound: Of Matter and Forces in the Physical World*. New York, NY: Oxford University Press.

Pasternack, S. (1938). 'Note on the Fine Structure of Ha and Da'. In: *Physical Review* 54.12 (Dec. 1938), p. 1113.

Pauli, W. (1933). 'Die allgemeinen Prinzipien der Wellenmechanik'. In: *Handbuch der Physik. Vol. 24: Quantentheorie*. Ed. by H. Geiger and K. Scheel. Berlin: Springer, pp. 83–272.

Pauli, W., A. Hermann and K. von Meyenn (1979). *Wissenschaftlicher Briefwechsel mit Bohr, Einstein, Heisenberg, Ua: Scientific Correspondence with Bohr, Einstein, Heisenberg, A. o.* New York, NY: Springer.

Peskin, M. E. and D. V. Schroeder (1995). *An Introduction to Quantum Field Theory*. Boulder, CO: Westview Press.

Roqué, X. (1992). 'Møller Scattering: A Neglected Application of Early Quantum Electrodynamics'. In: *Archive for History of Exact Sciences* 44.3 (Sept. 1992), pp. 197–264.

Sakurai, J. J. (1967). *Advanced Quantum Mechanics*. Fifth printing, Jan. 1976. Reading, MA: Addison-Wesley.

Sauer, T. (2008). *Remarks on the Origin of Path Integration: Einstein and Feynman*. URL: http://arxiv.org/abs/0801.1654.

Schönberg, M. (1946). 'Classical Theory of the Point Electron'. In: *Physical Review* 69.5–6 (Mar. 1946), pp. 211–224.

Schrödinger, E. (1930). 'Über die kräftefreie Bewegung in der relativistischen Quantenmechanik'. In: *Sonderausgabe aus den Sitzungsberichten der Preussischen Akademie der Wissenschaften, Phys. Math. Klasse* 24. Berlin: Verlag der Akademie derWissenschaften in Kommission beiWalter de Gruyter u. Co. Reprinted in Schrödinger 1984, pp. 357–368, 418–428.

Schrödinger, E. (1984). *Collected Papers*. Vol. 3. Vienna: Austrian Academy of Sciences.

Schwarzschild, K. (1903). 'Zur Elektrodynamik. I. Zwei Formen des Prinzips der kleinsten Action in der Elektronentheorie'. In: *Königliche Gesellschaft der Wissenschaften zu Göttingen. Mathematisch-physikalische Klasse. Nachrichten*, Issue 3, pp. 126–131.

Schweber, S. S. (1986a). 'Feynman and the Visualization of Space-Time Processes'. In: *Reviews of Modern Physics* 58.2, pp. 449–508.

Schweber, S. S. (1986b). 'Shelter Island, Pocono, and Oldstone: The Emergence of American Quantum Electrodynamics after World War II'. In: *Osiris. 2nd ser.* 2, pp. 265–302.

Schweber, S. S. (1994). *QED and the Men Who Made It: Dyson, Feynman, Schwinger, and Tomonaga*. Princeton: Princeton University Press.

Schwinger, J. (1948a). 'On Quantum-Electrodynamics and the Magnetic Moment of the Electron'. In: *Physical Review* 73.4 (Feb. 1948), pp. 416–417.

Schwinger, J. (1948b). 'Quantum Electrodynamics. I. A Covariant Formulation'. In: *Physical Review* 74.10 (Nov. 1948), pp. 1439–1461.

Schwinger, J. (1958). *Selected Papers on Quantum Electrodynamics*. New York, NY: Dover Publications.

Schwinger, J. (1983). 'Renormalization Theory of Quantum Electrodynamics: An Individual View'. In: *The Birth of Particle Physics*. Ed. by L. Brown and L. Hoddeson. Cambridge: Cambridge University Press, pp. 329–353.

Shapere, D. (1984). 'Notes Toward a Post-Positivistic Interpretation of Science, Part II'. In: *Reason and the Search for Knowledge*. Ed. by D. Shapere. Dordrecht: Reidel, pp. 352–382.

Shin, S.-J. (1994). *The Logical Status of Diagrams*. Cambridge: Cambridge University Press.

Stueckelberg, E. C. G. (1934). 'Relativistisch invariante Störungstheorie des Diracschen Elektrons I. Teil: Streustrahlung und Bremsstrahlung'. In: *Annalen der Physik* 413.4, pp. 367–389.

Stueckelberg, E. C. G. (1938). 'DieWechselwirkungs Kraefte in der Elektrodynamik und in der Feldtheorie der Kernkraefte (Teil I)'. In: *Helvetica Physica Acta* 11, pp. 225–244.

Stueckelberg, E. C. G. (1941). 'Remarque sur la création de paires de particules en théorie de relativité'. In: *Helvetica Physica Acta* 14, pp. 588–594.

Tödtli, B. (2004). 'QCD Corrections to the Wilson Coefficients for $b \rightarrow sl^{+}l^{-}$ in 2HDMs: Photonic Contributions'. MA thesis. University of Bern.

Tetrode, H. (1922). 'Über den Wirkungszusammenhang der Welt. Eine Erweiterung der klassischen Dynamik.' In: *Zeitschrift für Physik* 10, pp. 317–328.

Tomonaga, S.-I. (1946). 'On a Relativistically Invariant Formulation of the Quantum Theory of Wave Fields'. In: *Progress of Theoretical Physics* 1 (Aug. 1946), pp. 27–42.

Tomonaga, S.-I. (1973). 'Development of Quantum Electrodynamics'. In: *The Physicist's Conception of Nature*. Ed. by J. Mehra. Copyright 1966 by the Nobel Foundation and Elsevier, Netherlands. D. Reidel. Chap. 19, pp. 404–412.

Wüthrich, A. (2007). 'Book review: "Drawing Theories Apart: The Dispersion of Feynman Diagrams in Postwar Physics. David Kaiser."' In: *Studies in History and Philosophy of Modern Physics* 38.3 (Sept. 2007), pp. 586–589.

Walton, K. (1990). *Mimesis as Make-Believe: On the Foundations of the Representational Arts*. Cambridge, MA: Harvard University Press.

Weinberg, S. (1977). 'The Search for Unity, Notes for a History of Quantum Field Theory'. In: *Daedalus* 106.4, pp. 17–35.

Weinberg, S. (1995). 'Historical Introduction'. In: *The Quantum Theory of Fields*. Cambridge: Cambridge University Press. Chap. 1, pp. 1–48.

Weiner, C. (1966a). 'Interview with Dr. Richard Feynman, March 4 to June 28, 1966, Vol. 2'. Niels Bohr Library & Archives, American Institute of Physics, College Park, MD.

Weiner, C. (1966b). 'Interview with Dr. Richard Feynman, March 4 to June 28, 1966, Vol. 1'. Niels Bohr Library & Archives, American Institute of Physics, College Park, MD.

Weingard, R. (1988). 'Virtual Particles and the Interpretation of Quantum Field Theory'. In: *Philosophical Foundations of Quantum Field Theory*. Ed. by H. R. Brown and R. Harré. Oxford: Clarendon Press, pp. 43–58.

Weisskopf, V. F. (1934). 'Über die Selbstenergie des Elektrons'. In: *Zeitschrift für Physik* 89, pp. 27–39.

Weisskopf, V. F. (1939). 'On the Self-Energy and the Electromagnetic Field of the Electron'. In: *Physical Review* 56.1 (July 1939), pp. 72–85.

Weisskopf, V. F. (1983). 'Growing up With Field Theory: The Development of Quantum Electrodynamics'. In: *The Birth of Particle Physics*. Ed. by L. M. Brown and L. Hoddeson. Cambridge: Cambridge University Press. Chap. 3, pp. 56–81.

Welton, T. (2007). 'Memories of Feynman'. In: *Physics Today* 60.2. Written 1983, pp. 46–52.

Wentzel, G. (1943). *Einführung in die Quantentheorie der Wellenfelder*. Wien: Franz Deuticke.

Wheeler, J. A. (1948). '1948 Pocono Conference on Physics'. Notes prepared informally and unofficially by J. A. Wheeler. American Institute of Physics, Niels Bohr Library.

Wheeler, J. A. and R. P. Feynman (1945). 'Interaction with the Absorber as the Mechanism of Radiation'. In: *Reviews of Modern Physics* 17.2–3 (Apr. 1945), pp. 157–181.

Wheeler, J. A. and R. P. Feynman (1949). 'Classical Electrodynamics in Terms of Direct Interparticle Action'. In: *Reviews of Modern Physics* 21.3 (July 1949), pp. 425–433.

Index

Kolibri 360 GmbH
Ferdinand-Jühlke-Straße 7
99095 Erfurt, Deutschland
produktsicherheit@kolibri360.de